Contents

List of figures

Introduction

To the teacher

This is an 'ideas' book for teachers and students of A and AS level Geography. Many of the AS and A-level syllabuses now require a fieldwork project or enquiry as part of the examination. Although most A and AS level students will have completed GCSE fieldwork, their advanced work needs to have greater depth and be more analytical.

For the student, the first question is 'What shall I do my fieldwork on?' This book is intended to provide some suggestions for geographical enquiries with the emphasis on collecting primary data. It covers a range of themes examined in most specifications and it sets out a selection of aims, key questions or hypotheses, which the student can extract and apply to a local area.

One of the keys to success in AS or A-level coursework is the choice of topic. Is it manageable in scale; is it analytical in character; does the difficulty of the topic match the student's ability? We all use our professional judgement to guide students towards work which will stretch them but remain within their grasp. This book identifies topics for investigation, outlining suggestions for data collection and techniques for recording and analysing data. Ideas are given for graphs, maps and statistical techniques that would be appropriate. Some of the ideas are simplistic, while others are more complex, because it is hoped that students of a range of abilities will find this book useful.

Discussions with students over choice of topic for a fieldwork project can be time-consuming and repetitive, as can the overall supervision of each project. The wide variety of fieldwork outlines here will help students on their way as independently as possible. Geography teachers are a fountain of good ideas to be tapped, and all have tried and tested fieldwork favourites. No 'ideas' book will ever be complete, and in a changing world a good geographer will always come up with something new. However, it is hoped that this book will save teachers a great deal of time, and give students the confidence to get started.

To the student

This book is designed to give you some ideas for the coursework part of your geography examination. You need to choose carefully so that you produce the best work you can. Beware of extending or revamping your GCSE fieldwork. Almost all students find it hard to return to work and look at it in greater depth, as required for AS and A2.

Think about what you are interested in and which parts of your geography syllabus you understand most fully. There are a number of titles described here. Select a broad topic which interests you and consider some of the specific ideas. You can take part or all of each idea, or combine them, depending on your location. Make sure though that the 'package' you adopt hangs together and that you can develop a simple aim or statement for which all your data collection is relevant. You should adapt the ideas in this book to suit your own locality.

You may decide to make a comparative study – of two rivers, two football clubs or two leisure centres, for example. Remember to state the *reasons* for such a comparison. There must be some geographical logic to your choice.

Think very carefully about the *scale* of your project. It must not be so small that you won't collect sufficient data to reach conclusions. Likewise you should not take on something so large that you cannot collect all the data you need in the time available.

Think also about practicalities. Can you visit a particular area easily? How much time do you have to complete the work? Are you well organised? Do you wish to work with a friend? Is he/she reliable? How will you ensure that your work is individual? When is the best time of year to complete your fieldwork? If you are working on vegetation it is easier when plants are in flower in the summer. Similarly, measuring pebbles in a river can be miserable in winter when the days are cold and wet.

How to use this book

Decide on your area of interest then read up all the ideas on that particular broad topic. The titles are there to start you thinking and planning, and need not be prescriptive. You may well find that some of the ideas under one title can be used in another. Similarly references and resources from one title may be useful in another.

FIELDWORK IDEAS in ACTION

Gill Miller

Hodder & Stoughton

A MEMBER OF THE HODDER HEADLINE GROUP

Dedication and Acknowledgements

To Rory, Sarah and Aidan

Thanks are due to Rory Miller and Chris Lane for their invaluable advice on reading the manuscript, and to all those students and colleagues whose fieldwork experiences provided the inspiration for this book.

The publishers would like to thank the Field Studies Council of the Slapton Ley Field Centre for permission to use figure 24 on page 33 (soil analysis).

Every effort has been made to trace and acknowledge ownership of copyright. The publishers will be glad to make suitable arrangements with any copyright holders whom it has not been possible to contact.

Orders: please contact Bookpoint Ltd, 78 Milton Park, Abingdon, Oxon OX14 4TD. Telephone: (44) 01235 827720, Fax: (44) 01235 400454. Lines are open from 9.00–6.00, Monday to Saturday, with a 24 hour message answering service. Email address: orders@bookpoint.co.uk

British Library Cataloguing in Publication Data
A catalogue record for this title is available from The British Library

ISBN 0 340 75355 2

First published 2000
Impression number 10 9 8 7 6 5 4 3 2
Year   2005 2004 2003 2002 2001 2000

Typeset by Wearset, Boldon, Tyne and Wear.
Printed in Great Britain for Hodder & Stoughton Educational, a division of Hodder Headline Plc, 338 Euston Road, London NW1 3BH by Hobbs the Printers Ltd.

A plan of campaign to complete your fieldwork enquiry

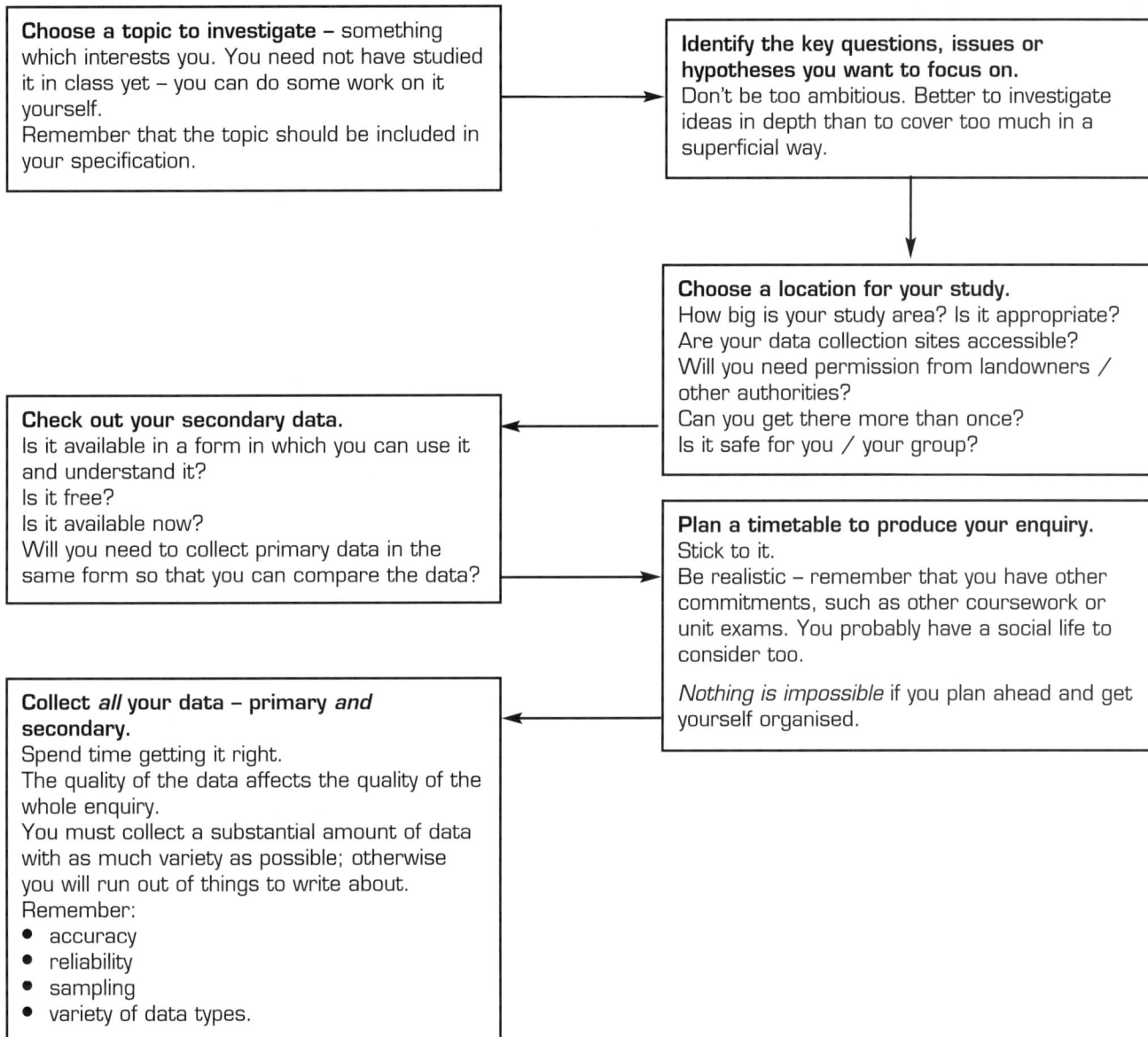

Choose a topic to investigate – something which interests you. You need not have studied it in class yet – you can do some work on it yourself.
Remember that the topic should be included in your specification.

Identify the key questions, issues or hypotheses you want to focus on.
Don't be too ambitious. Better to investigate ideas in depth than to cover too much in a superficial way.

Choose a location for your study.
How big is your study area? Is it appropriate?
Are your data collection sites accessible?
Will you need permission from landowners / other authorities?
Can you get there more than once?
Is it safe for you / your group?

Check out your secondary data.
Is it available in a form in which you can use it and understand it?
Is it free?
Is it available now?
Will you need to collect primary data in the same form so that you can compare the data?

Plan a timetable to produce your enquiry.
Stick to it.
Be realistic – remember that you have other commitments, such as other coursework or unit exams. You probably have a social life to consider too.

Nothing is impossible if you plan ahead and get yourself organised.

Collect *all* your data – primary *and* secondary.
Spend time getting it right.
The quality of the data affects the quality of the whole enquiry.
You must collect a substantial amount of data with as much variety as possible; otherwise you will run out of things to write about.
Remember:
- accuracy
- reliability
- sampling
- variety of data types.

Primary Data

Give yourself plenty of time – data collection always takes longer than you think. If you discover that your data collection is not viable, then you have time to change your ideas or amend your title. Your **pilot study** is really important.

Secondary Data

If you want data from other people, such as local authority planners, Environment Agency or private businesses, you must be courteous and organised. Some organisations receive many requests for information from students of all ages and it can become very tiresome. It is therefore important to be very specific about what you are doing and why, and the data you require. There is no point in sending a letter to the Planning department saying 'I am doing a project on the CBD. Can you give me some information?'.

Writing up your enquiry

Get the mark scheme.
Be sure you know what you need to do to get high marks. Read the mark descriptions carefully – they will tell you what your work has to be like to reach certain marks. Use the mark scheme as a guide to what is expected of you.

If you can word-process your enquiry, so much the better. It will be easier to read, and much easier for you to correct and amend. You will also be more satisfied with the end result if it looks good.

Remember – *do not have regrets at the end of the day, thinking that you could have done better.*

Here are some tips to follow

- It is easier for the examiner if you follow the structure of the mark scheme in your writing up.
- State your **aim** clearly and list any hypotheses you will investigate. There should be some logical order to your hypotheses so that the whole coursework hangs together.
- Explain briefly the geographical background to these ideas. It is not always necessary to state what you expect to find. If this is genuine research, you may not know.
- Give a short description of your study area and locate it clearly. That does **not** mean a British Isles map with a dot on it! Draw two maps – one regional and one local – to put the study into context. Only put information on your maps that is relevant to the topic. If you are investigating a river, for instance, topography and land use are important. Details of picnic sites, road networks and specific buildings will only clutter the map. Likewise, an investigation into urban regeneration does not need every detail of topography and every stream. **Do not** photocopy maps from road atlases, which have lots of superfluous information on them, and worse still **do not** enlarge or reduce maps without paying attention to the **scale**. Every map needs a scale and all geographers should be able to work one out.
- Explain your data collection methods carefully. Avoid being too descriptive – you don't need to draw pictures of soil augers or clinometers, or include photographs of a ranging pole. You **do** need to justify your sampling methods and explain why they are suitable for their purpose. Good geographers visit their study sites before they begin their research and conduct pilot exercises to test questionnaires (see figure 1) and equipment, and to confirm sampling sites. Be aware of the **limitations** of your data collection methods. Do not rely wholly on questionnaires. Collect other data too.
- Beware – your evaluation at the end should not be a list of excuses for things that went wrong. Most difficulties should be ironed out at the **pilot stage**. You will give a good impression if you recognise a problem and do something about it. If word limit is an issue in your specification, consider summarising your data collection in tabular form.
- When you analyse your results try to integrate all the information, graphs and photographs together within your text. It is not very effective to place all the graphs or all the photographs in a separate section at the end. You will find it easier to use all the material relating to a particular hypothesis or question if it is located together. Aim for variety of presentation of data. Avoid the 'death by pie charts' approach more typical of GCSE. Every item should have some annotation to show why it is included in the study. Graphs should have a comment about the general pattern shown plus any anomalies. Photographs need more than a label such as 'the village shop' or 'stream site 1' to guide the examiner to look in the correct context. Make sure that all photographs have a purpose. Similarly, maps need amplified comment on the patterns shown. Remember to refer to every resource in your writing up.
- Try to complete your personal observations first, so that you can raise queries or issues with the people you interview, if necessary. You can always return to your fieldwork sites for further observation but it is very unlikely that you will be able to interview someone twice.
- Make sure that you do not *describe* your data instead of *analysing* it. Your text needs to draw out relevant key features and then begin to *explain* the results to address your original aim.
- In the concluding section you should refer systematically to your original aim / hypotheses / statements and comment on each one in turn. Have you done what you set out to do? Can you summarise your findings? Are they what you expected? Are there unusual results which you could investigate further?
- Try to evaluate your work. Are your results typical of theoretical or textbook information? If not, why not? How could your work develop further? What other avenues of research could be followed? Where do your results fit into the geography of the local area? Above all, avoid a list of 'could have done better if I'd planned this / spent more time / worked harder'. This is particularly important if you don't actually do what you set out to do.
- Check the specification rules regarding appendices. In general there should be no work in an appendix which is central to your study but you may wish to include a sample of a completed questionnaire. Do not include tourist brochures and leaflets which you have accumulated during your work. If they have a purpose, add them to the main body of your coursework.
- Finally, include a bibliography. This should not be a list of well-known geography textbooks. Never include material in your study without acknowledging it in the bibliography.

Figure 1 Questionnaires

Sample size is always difficult. In a large town a sample of 100 is tiny, but at a corner shop you could wait a very long time for 100 people. Consider size but also *how long* it is likely to take you to collect a given number of responses. You must justify your sample size.

The design of your questionnaire is very important. A poor one will give inadequate data which will result in poor coursework, but a good questionnaire can provide a substantial base for analysis.

Consider the following:
- Balance the number of questions with the time taken to answer them. Most people will not want to be held up for long. However, make sure you get what you need to know in the form in which you need it.
- Use closed questions with a selection of possible answers, eg daily / weekly / fortnightly / monthly / other.
- Use scaled answers where respondents ring their answer on a sliding scale, eg from 1 to 5.
- Do not ask personal questions. Either estimate age or ask respondents to identify an age group, eg under 21, 22–40, 41–60, over 60.

Dos and Don'ts

DON'T	DO
Don't keep putting off getting started. You will enjoy the challenge of doing something independently.	**Do** beware of topics that are **comparisons** – there must be some 'geographical logic' to a comparison. Why *should* there be differences? Places need to have *some* similarities so that you can focus on the significant differences.
Don't assume that factory managers, shopkeepers, hoteliers, police, etc will have or will be willing to give you data. Don't bother other people for data if you don't know precisely what you want. You'll look foolish and ruin the chances and opportunities for the next student.	**Do** collect lots of different types of data. The more you have the more you can analyse.
Don't make unreasonable or unsubstantiated claims about your results. At this level students should appreciate that results may only be partial, incomplete or even tentative. It is by no means seen as a 'failure'.	**Do** remember that consistency in your data collection is very important. You may not have access to expensive equipment but you can still identify patterns if you work at every sampling site in the same way.
Don't wish that you had spent more time, collected data more carefully, thought about the results in greater detail. Some examination boards will allow you to resubmit your enquiry, but you will probably have had enough of it by then. Be determined to do your very best from the beginning and don't have any regrets when you get your results.	**Do** remember that your data can never be 'wrong'. If you have measured it carefully and honestly and if doesn't 'fit' your predicted pattern, ask yourself why not. Think about your method of sampling, or the location, or other geographical factors that could affect your results.
	Do ask for advice from teachers – they know what is needed and have a wealth of expertise. But they won't do the work for you. You will have to have something on paper to discuss with them.
	Do meet your deadlines. You will feel much more confident when you do. This is a must. The exam boards will not wait for you and your teachers will run out of patience if you are continually late with work. Time has a horrible habit of catching up with you. If you miss a deadline it just means more pressure on the next one. Be warned!

How do hydraulic variables change downstream?

Starting Points ▶

1. Width and depth of a river increase downstream.
2. Discharge increases downstream.
3. Streams become more efficient downstream.
4. Investigate the interrelationships between these variables downstream.
5. Geology and land use affect hydraulic variables along a river.
6. Urbanisation / water abstraction have an impact on river variables.

Geographical links to your syllabus

- Streams exhibit dynamic equilibrium, ie a balance between hydraulic variables.

- Changes in geology, slope and land use affect the hydrological cycle.

- There are many interrelationships between variables affecting the flow of a river.

Primary Data Collection

1. Select at least 10 sites along your stream or river. Select your sites carefully depending on the emphasis of your study, eg if geology is important, decide where to locate your sites to minimise any changes due to other factors. If you intend to look at the impact of urbanisation, choose a balance of sites in urban and non-urban locations.
2. Measure width of water at each site.
3. Measure depth of water at regular intervals across the river – 10 or 20cm intervals.

⚠ *Things to look out for*

The selection of your sites is critical. You must be **safe** – do not work in a river which is clearly too deep or fast flowing.

You must choose sites which are some distance apart. You will not find much change in rivers along a few hundred metres and you will struggle to find sensible comments to make. An investigation over 10km or more will be much better.

Check the access of every site – do not trespass. Most farmers will give permission if they understand what you are doing.

Make a pilot study to test your data collection methods. Make sure that your sites do actually have water in them!

Time for your fieldwork is important. You must collect as much data as you can in one day, and preferably all of it. Some rivers have a rapid response time to rainfall events and you could find it difficult to make sensible conclusions if the environmental conditions change dramatically.

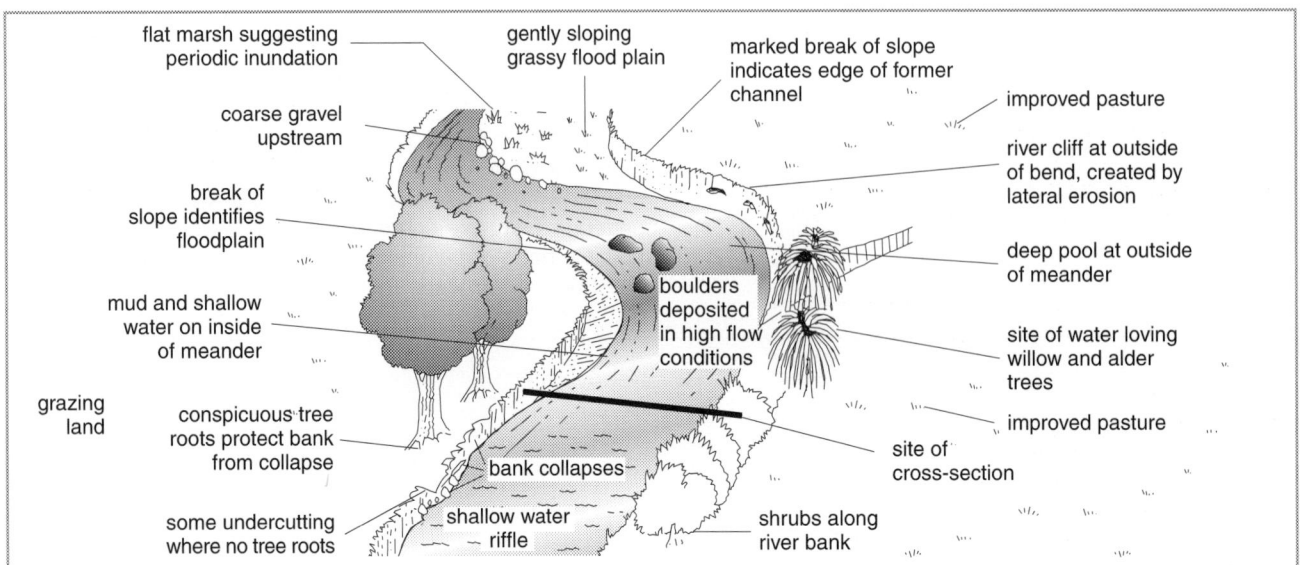

Figure 2 Annotated sketch to show the characteristics of a river site

4. Measure velocity through each site. Average of three readings – time taken for an orange to flow 10m. (Remember that the orange should be moving before it reaches the point at which you start timing it.) If you use a flow meter calculate the depth of the impeller (0.8 of depth from the bed), and select your site in the stream carefully. You may decide to take the average of several velocity readings across your river.

5. Measure the wetted perimeter by stretching a chain or tape over the sides and bed of the channel. This will give the actual length of 'wetness' in the channel.

6. Measure the slope of the stream bed going through your measuring site – angle over at least 10m.

7. Make a detailed description of each site – undercutting, vegetation, channel pattern, calibre (size) of bedload. Annotate a sketch diagram to indicate characteristics which may have an effect on your results. (see figure 2).

8. Map the land use along your river.

Secondary Data Collection

1. Map the geology along your river.
2. Find out about water abstraction from your river.
3. In an urbanised area, collect details of river engineering schemes to manage flow.
4. If your stream is monitored by the Institute of Hydrology collect recent data.

Ideas for recording and analysing data

1. Maps. Details of study area with key features such as location of sites, lengths of managed river, built-up areas, sites of water abstraction. Geology map. Land use map.
   Do not be tempted to put too much information on one map.

2. Annotated sketch diagrams / photographs / large-scale sketch maps of each site to set the scene for each set of results.

3. Annotated river cross sections. Put several on a page, under one another, using the same scale. This means you will be able to compare size more easily and any unusual changes can be clearly identified. Do not exaggerate the vertical scale. Your sections will look very shallow – but, after all, the water was probably only knee deep.

4. Calculate discharge in cubic metres / second (cumecs). Use average depth × width × velocity. Alternatively, you can calculate channel area from your accurate cross sections and then multiply by velocity.

5. Scatter graphs of the relationships you are investigating. Several on one page will help you to identify interrelationships and especially anomalies at particular sites. You may decide to draw in best fit lines to summarise the relationships which are shown. Beware the default line on a computer graph. It will respond to the data shown and may not identify a geographical relationship.
   Remember to annotate. Pick out any unusual results. Are there similar anomalies shown on any of the other scatter graphs?
   Examples of scatter graphs: width v distance downstream; depth v distance downstream; velocity v distance downstream; discharge v distance downstream; hydraulic radius v distance downstream; interrelationships such as width / depth ratio v distance downstream; velocity v discharge.

STATISTICAL ANALYSIS
If you have at least 10 sites along your river use Spearman Rank correlation to test the strength of the relationships you have identified. You should also test the r_s value for significance – how reliable is that relationship?

Interpretation and Conclusions

- How do different variables change downstream?
- If there are changes, why do they occur? If not, why not?
- Does your river change as textbooks suggest? How? Why? Can you claim from the statistical results that your river is typical?
- Are there particular sites along your river which show unusual characteristics? Try to explain them.
- How does land use affect river variables? Does geology have an influence? How? What evidence is there? How does urbanisation affect river variables?
- How do your results compare with those from the Institute of Hydrology? What are the differences and why do they occur?
- Assess the quality of your results in the light of your analysis. *Do not make this an excuse for poor data collection.* Are there ways in which your data collection could be improved in the light of your findings? Will season have an influence on your results?
- Can you make some generalisations about river variables which could be investigated further, perhaps on other rivers or streams?

Resources

Institute of Hydrology: Maclean Building, Crowmarsh Gifford, Wallingford, Oxfordshire OX10 8BB. Tel: (01491) 838800. Fax: (01491) 692424. Web site: www.nwl.ac.uk/ih (click on 'Water watch')
River Data Centre website: www.nwl.ac.uk/requesting-data.htm
Environment Agency website: www.highdown.berks.sch.uk/hydrology
Streamwatch website: www.mmu.ac.uk/c-a/edu/streamwatch

Analysis of interrelationships between channel efficiency and river bedload

Starting Points ▶

1. Channels become more efficient downstream.
2. Inefficient channels are characterised by coarse and poorly sorted bedload material.
3. Bedload in rivers is more closely related to bankful discharge than to low level flows.
4. Channel cross-sectional area is inversely related to bedload size.

Geographical links to your syllabus

- Erosive power of rivers is greatest at bankfull, when the largest material can be transported. In discharges less than bankfull, deposition of largest material is likely.

- Bedload material becomes smaller or finer through continued attrition downstream.

- Hydraulic radius, a measure of channel efficiency, is greatest in channels carrying fine material.

- Higher discharges downstream carry more bedload. Capacity and competence increase.

Primary Data Collection

1. Select 10 sites at least one kilometre apart along a river. Alternatively choose three or four sites on several neighbouring streams. Try to select similar sites, eg all on straight reaches, all pools, or all riffles. Try to eliminate more variables by choosing streams on similar geology.
2. At each site measure wetted perimeter, width of water, depth of water at regular intervals across the river, velocity through the site.
3. Measure bankfull cross section at each site. Stretch a tape across the width of the channel from bank to bank. Measure depth from the horizontal tape (flood plain) to bed of the river (see figure 3).
4. Measure the slope of the river bed through each site, and the slope of the flood plain.
5. Sample 20–30 pebbles across each site and record length of 'a' axis. Think about your sampling method. Use a systematic sample every 10 or 20cm across the width or use random number tables to identify locations along the tape stretched across from bank to bank. This sample should include material from the river *channel*, not just the section under water. Record the location of each pebble.
6. Make a detailed description of each site to include land use, character of the channel, nature of banks, evidence of bank collapse, overhanging vegetation. Include sketches and / or photographs.

⚠ *Things to look out for*

Your personal safety is a priority.

This topic works best on streams with coarse bedload rather than fine silts and clay.

Justify your sample size and your sampling method.

If you are working on several tributaries, make sure they are as comparable as possible, ie similar geology, land use, rainfall.

Secondary Data Collection

If you work on a monitored river you may find data from the Institute of Hydrology for bankfull discharges. You should try to match your sampling sites with those used by the Institute of Hydrology.

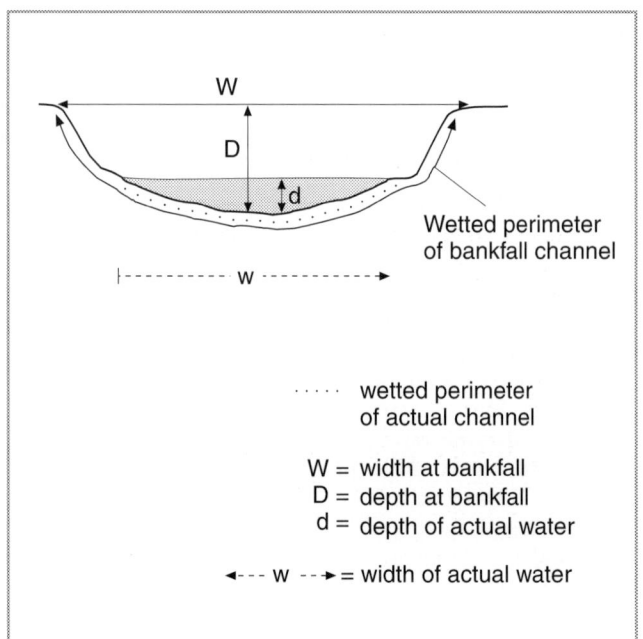

Figure 3 Distinguish between bankfull and river cross sections

Ideas for recording and analysing data ▶▶

1. Calculate the hydraulic radius for each site, using data for the actual water channel. [HR = Area of cross section / wetted perimeter]
2. Calculate the hydraulic radius (HR) of the bankfull channel in the same way, using measurements of the whole channel.
3. If you worked on only one stream, draw a scatter graph of HR against distance downstream. Use Spearman Rank to correlate these two variables. Don't forget to test for significance. How reliable is your relationship?
4. Construct a dispersion diagram for each site (see figure 4) to graph 'a' axis of pebbles from the river itself (not the dry part of the channel). Calculate the range of values, mean, median, interquartile range, standard deviation. How does standard deviation from each site change?
5. Repeat the dispersion diagram using *all* the pebble data from each site. Is there a difference between the bankfull situation and low flow conditions? Remember to annotate every graph to summarise its meaning and to identify any anomalies.
6. Scatter graph of cross-sectional area *v* mean bedload size. If you are able to draw a best fit line, consider a Spearman Rank correlation to find if there is a relationship between them.
7. Calculate actual discharge and bankfull discharge for each site. Draw graphs to illustrate the relationships of actual discharge *v* bedload size (measured as the 'a' axis) and bankfull discharge *v* bedload size.

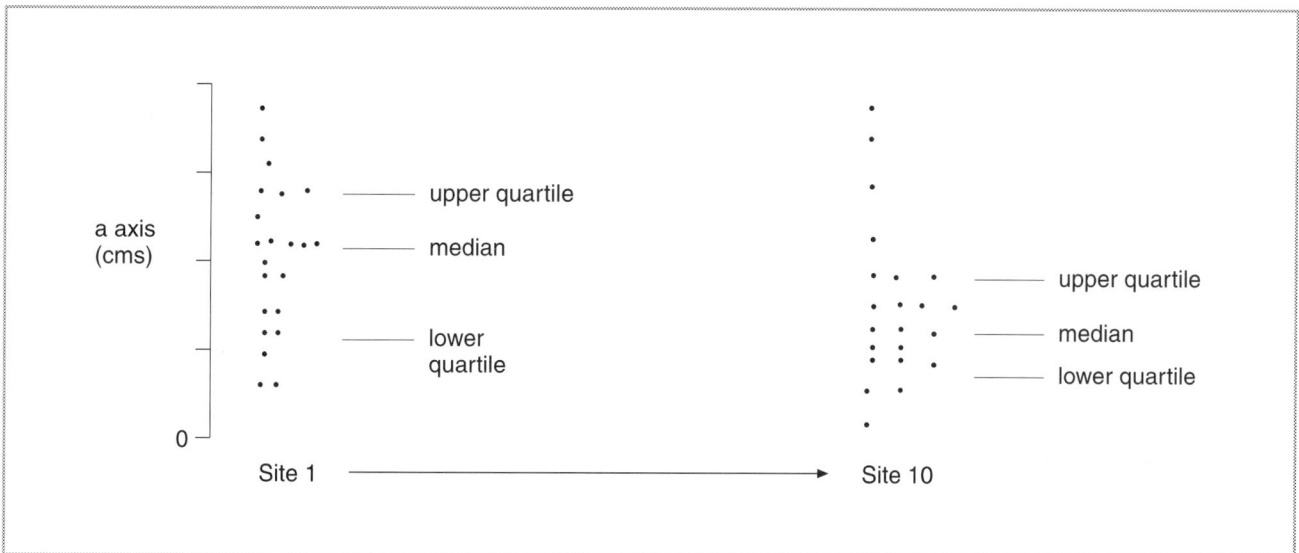

Figure 4 Dispersion diagrams represent range of bedload sizes

Figure 5 Calculate bankfull discharge

$Q = W \times D \times$ bankfull velocity.
Estimate bankfull velocity using Mannings 'n'.

$$V = \frac{R^{0.67} \times S^{0.5}}{n.}$$

R = Hydraulic Radius
S = slope of floodplain
n = Mannings 'n'.

Estimated figures for 'n' can be found in D Waugh: *Geography an Integrated Approach* and Clowes & Comfort *Process and Landform*.

Interpretation and Conclusions

- To what extent does your river become more efficient downstream? Are there any exceptional sites? From your personal observations / sketches can you suggest why? If there is no change in efficiency can you explain why not?
- Do inefficient channels have a greater range of bedload size? Where? Why?
- Compare the differences in relationships at bankfull and in low flow conditions.

Resources

'Measuring Upland Streams', *Geography Review*, Volume 11, Number 1, September 1997
Hydrological Data Yearbook from the Institute of Hydrology at Wallingford; shows daily, monthly flows and flow duration statistics.
Website: www.nwl.ac.uk/ih
Meteorological Office website: www:meto.govt.uk

An investigation of changes in suspended load and bedload downstream

Starting Points ▶

1. Suspended load is related to geology and land use.
2. Suspended load increases with discharge.
3. Bedload shape changes downstream.
4. Bedload becomes smaller and more rounded downstream.
5. There is a relationship between bedload size and bankfull cross sections.
6. The volume of suspended load and bedload can be related to erosion along river banks.

Geographical links to your syllabus

- A variety of processes of erosion act within channels.
- Bankfull discharges have more energy therefore result in most erosion.
- Size of bedload is related to velocity (refer to the Hjulström graph in your textbooks).
- Land use such as unvegetated surfaces or deforestation may lead to high levels of suspended load.

Primary Data Collection

1. Select at least 10 sites downstream, spread over a substantial distance – at least 10km. Record a detailed description of each site with annotated sketches to identify important characteristics such as channel pattern, land use on flood plain, constrictions to banks of river etc.
2. At each site measure the river cross section by securing a tape across the river at the top of each bank. Measure width, depth (at 20cm intervals) and velocity of actual water in the stream.
3. Select at least 20 pebbles systematically from across the river bed. Measure a, b and c axes of each pebble.
4. To give an indication of suspended load, measure turbidity (cloudiness) of the water at each site using a Secchi disc (figure 6).
5. Collect a sample of suspended sediment load from each site (see figure 7). Label each one carefully in identical jam jars.
6. If you have access to a set of sieves you can collect a bedload sample to analyse in the laboratory.
7. Record land use along the length of river under investigation up to 100m from the river.

Secondary Data Collection

Obtain a geology map of the area.

- Take a 20cm diameter piece of wood and paint it white.
- Attach it to a rod calibrated in 5cm intervals.
- Put the disc on the river bed and raise it gently.
- Note the height from the river bed at which the white disc is first visible. The greater the distance from the river bed, the less the turbidity.

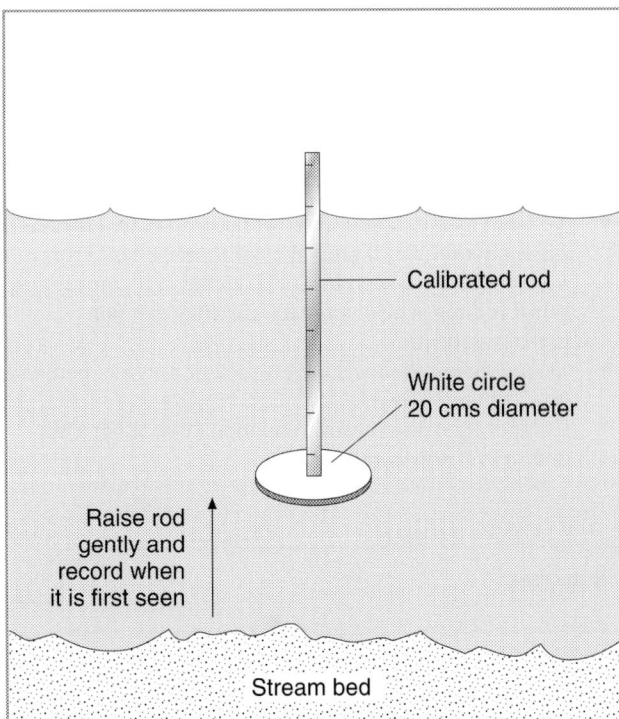

Calibrated rod

White circle 20 cms diameter

Raise rod gently and record when it is first seen

Stream bed

Figure 6 Measuring turbidity using a Secchi disc

⚠ Things to look out for

Your personal safety is paramount. Do not take risks in deep, fast-flowing rivers. Do not underestimate the force of flowing water.

Remember that the discharge you measure on your fieldwork day is not related to the amount and size of bedload in the river. River load is transported at very high discharges and deposited as discharge falls. You should not be working in high discharge environments.

This study works best in drainage basins of uniform geology. If your stream flows over different geology you could consider a comparative study to see how much underlying geology affects sediment transport.

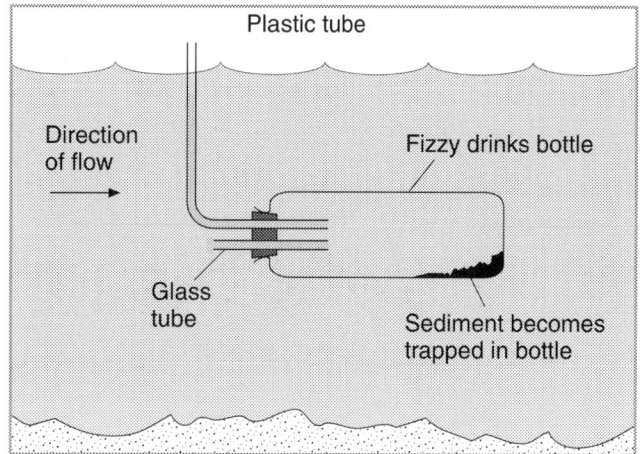

Figure 7 Use a plastic bottle to collect suspended sediment

Ideas for recording and analysing data   ▶▶

1. Draw an annotated map to show location of each site. Identify any particular features which may affect bedload and suspended load eg deforestation in progress, urban area, artificial banks built to contain the river or stream. Record in detail evidence of erosion along the banks of the river.
2. Map the land use along the river within 100m either side of the channel. Don't use too many categories. One example could be arable, pasture, rough grazing, waste land. Or residential, industrial, commercial.
3. Draw annotated sketches or take photographs that describe each site in detail. Focus on the evidence of erosion in particular.

To analyse suspended load:
1. Shake each identical jam jar of river water thoroughly and leave to stand for five minutes. Measure depth of settled sediment at the bottom of each jar (coarsest fraction). Repeat after 30 minutes (medium fraction) and after 24 hours (fine fraction). You can then compare the proportions of suspended load sizes at each site.
2. Draw divided bar graphs of each site to represent proportions of coarse, medium and fine sediment. Put as many as you can on one page so that you can compare them easily.
3. Calculate river discharge for each site. Draw a scatter graph of discharge against each component of suspended sediment load to investigate how suspended load changes with discharge downstream (see figure 8).
    Draw a best fit line to assess whether or not there may be a relationship between variables. Use Spearman Rank correlation to investigate the following:
    discharge *v* coarse suspended load
    discharge *v* fine suspended load.

To analyse bedload select from a number of ways of using the shape and size of material.
1. Draw dispersion diagrams for the a axis of the 20 pebbles from each site.
2. Identify mean, median, interquartile range of a axis or calculate standard deviation for each site.
3. Calculate Krumbein's index of sphericity (see figure 9) or use the Cailleux index of roundness.
4. Use the a, b, and c axes to classify pebble shape on a Zingg diagram (see figure 10). Compare the shapes at different sites on your river, perhaps the first and last site, to see how / if shape changes downstream. Use Chi Squared to investigate whether there is a significant difference in the numbers of each shape at different sites.
5. Draw scatter graphs to investigate changing relationships downstream, eg mean roundness *v* distance downstream; a axis *v* distance downstream. Make sure you can justify any relationships you graph, annotate each graph, and look for explanations of the pattern of your results.
6. If you are able to make a sieve analysis, dry each sample and note the total weight. Weigh the proportions in each sieve size and calculate the percentage of total weight in each sieve size. Compare different sites, perhaps the furthest upstream and downstream, to see if there is a significant difference between bedload size at each site. Draw frequency curves to illustrate the difference in the proportions of bedload sizes at different sites (see figure 11). The Wentworth Scale (see figure 12) describes the sizes of sediments.

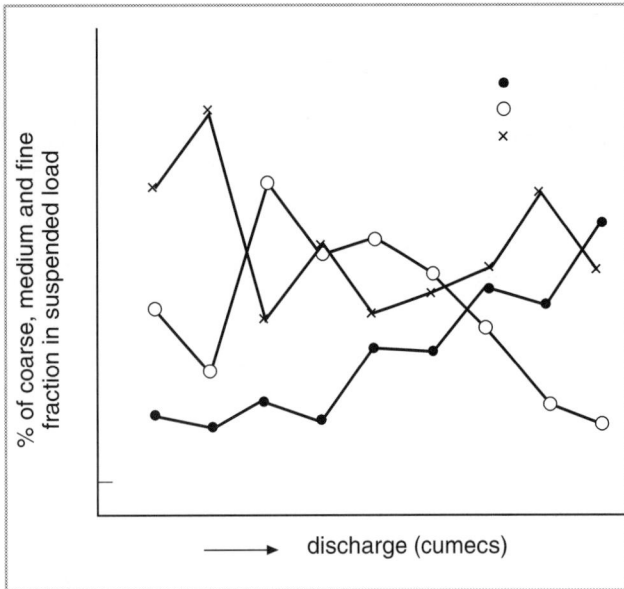

Figure 8 Changes in calibre of suspended sediment with discharge downstream

(y-axis: % of coarse, medium and fine fraction in suspended load; x-axis: discharge (cumecs))

Krumbein's sphericity Index.

$$K = \sqrt[3]{\frac{bc}{a^2}}$$

A perfect sphere = 1.0

Cailleux Index of Roundness

$$R = \frac{2r}{a} \times 1000$$

where r = roundness of the sharpest corner.
A perfectly round pebble has an index of 1000.

Figure 9 Krumbein and Cailleux

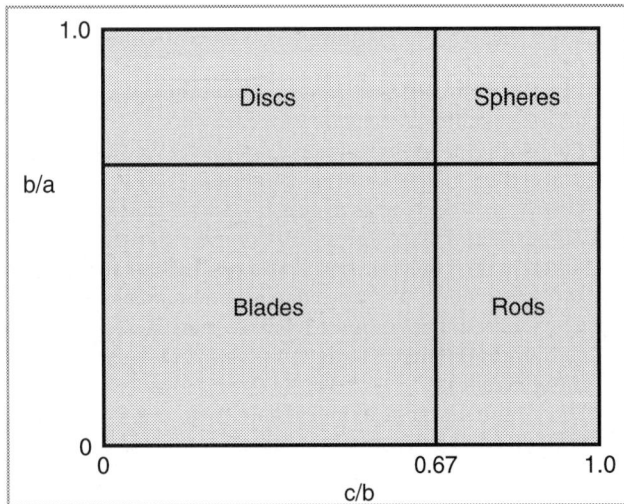

Figure 10 Zingg diagram

(Zingg diagram: b/a versus c/b with regions Discs, Spheres, Blades, Rods; c/b axis marked 0, 0.67, 1.0; b/a axis marked 0, 1.0)

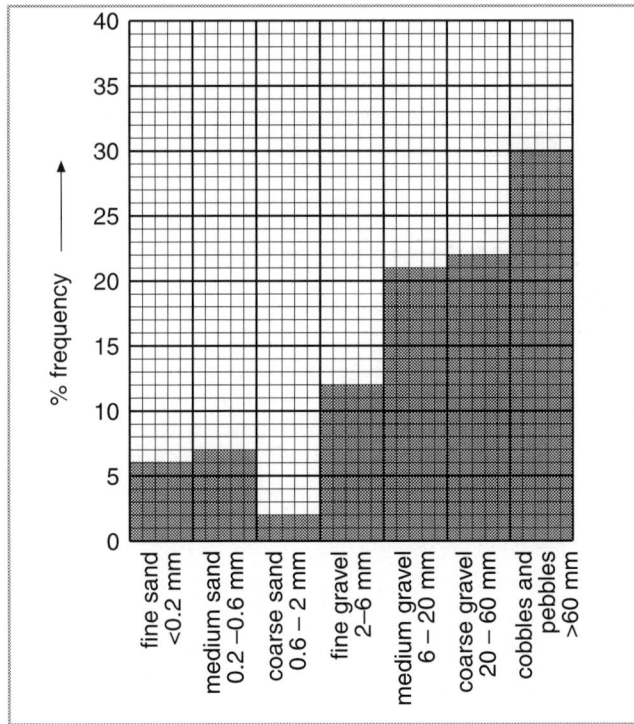

Figure 11 Frequency curve to show proportion of bedload

(% frequency versus size classes: fine sand <0.2 mm, medium sand 0.2–0.6 mm, coarse sand 0.6–2 mm, fine gravel 2–6 mm, medium gravel 6–20 mm, coarse gravel 20–60 mm, cobbles and pebbles >60 mm)

Figure 12 Wentworth Scale – standard size scales used to describe soils and sediments

Descriptive term	Size range
boulders	over 250mm
cobbles	60–250mm
coarse gravel	20–60mm
medium gravel	6–20mm
fine gravel	2–6mm
coarse sand	0.6–2mm
medium sand	0.2–06mm
	(250–500μm)
fine sand	0.06–0.2mm
	(62–25μm)
silt	2–62μm
clay	less than 2μm

μm = micrometre (or micron)
    = 0.001mm

Interpretation and Conclusions

- How do the suspended sediment fractions change downstream? Why should this happen in your particular river? Is there any link with erosion of river banks?
- Can any changes observed be linked to land use, geology or man's activities?
- Is there a relationship between discharge and amount of suspended load? Can you explain your answer?
- To what extent do particle roundness and size change downstream?
- What other factors influence roundness and bedload at any particular time?

Resources

Environment Agency website: www.environment-agency.gov.uk/
National Rivers Authority website: www. open.gov.uk/nao

What is the relationship between channel pattern, bedload and stream efficiency?

This is a very focused study looking at bedload in different channel patterns. If you make a comparison, make sure that there is at least one element which is similar. You could look at similar channel patterns on different geology or on different streams. Alternatively, you could contrast channel patterns on the same geology; or similar patterns in the upland and lowland reaches of the same stream if geology is similar.

Starting Points ▶

1. Channels with a similar pattern will contain similar bedload characteristics.
2. Channels with different patterns will have contrasting bedload characteristics.
3. Channels with different patterns and bedload characteristics reflect differences in stream efficiency.

Geographical links to your syllabus

- Meandering, braided and straight channels often have distinctive and contrasting shapes.

- Channel shape affects stream energy which in turn influences size of bedload within the channel.

- The efficiency of a channel, measured as the hydraulic radius, reflects stream energy, steam velocity and consequently size of bedload deposited in the channel.

Primary Data Collection

Consider carefully what channel patterns are present in your chosen stream. You may choose to investigate contrasts, eg between meandering and straight reaches.

1. Select at least five meandering stretches (reaches) of river, five braided reaches and five straight reaches along the same stream.
2. At each site survey the river cross section. (It is the *bankfull stage* which is most relevant here because bedload of a given size is transported when the stream has maximum energy, ie at bankfull.) Stretch a tape across the river joining the top of each bank. Record channel width. Measure the depth of the channel from the tape to channel bed at regular intervals. (Do not measure the depth of the water.) Measure the wetted perimeter at the *bankfull stage* (see figure 3).
3. Observe each site carefully to comment on the roughness of the channel. Record such influencing factors as vegetation, logs, sand bars, channel irregularities, incorporating a visual impression of the size and shape of bedload.
4. Take a systematic sample of bedload across the river. Measure a, b and c axes and roundness (use the Cailleux index of roundness, or Powers roundness chart). Select **at least** 20 pebbles.

5. If the bedload is too fine to measure, collect samples in an ice cream tub for a sieve analysis. Take samples from near the banks *and* in the centre of the stream.
6. To analyse the fine material, dry the sample thoroughly. Mix the sample with a spoon then separate 200gm into a set of sieves. After shaking vigorously for 20 minutes weigh the fractions in each sieve. Calculate the percentage of each particle size in the sample.

Secondary Data Collection

Obtain a large-scale OS map and a geology map of the area.

⚠ *Things to look out for*

You must be sure that you have safe access to several sites along the river.

You need as many sites as possible. Remember your emphasis for *comparison* – different patterns, different geology or upland/lowland.

Ideas for recording and analysing data

Remember that the emphasis is on comparing the categories of channel pattern, so group the diagrams of your sections of stream together.

For each group of channels:
1. Draw a detailed cross section of each site. Do not exaggerate the vertical scale too much – it doesn't have to be huge. If they are all on one page you can compare sites easily.
2. Draw a scatter graph of a axis values from the outside of the meander to the inside. Use different colours for each meander you have studied. Is there a pattern?
3. Draw similar graphs for roundness values across the river at each site.

4. If you used a sieve analysis, draw a divided bar chart of results for each site. Compare differences between sites with similar patterns. Try to summarise results for each channel pattern then compare results *between* channel patterns.
5. Calculate the hydraulic radius at bankfull for each site you studied. For each group of sites calculate the standard deviation of the hydraulic radius. This will enable you to identify variation between the different groups of channel patterns. The smaller the standard deviation the more similar the results within each group.
6. Compare the results for the different channel patterns.

Interpretation and Conclusions

- Summarise the channel shapes of each group of reaches you are working with. Describe how bedload size and roundness change across the channels.
- Discuss the relationship between channel efficiency, as measured by hydraulic radius, and channel pattern?
- Is there a relationship between hydraulic radius and bedload size as represented by mean a axis? Why?
- Establish the similarities and differences between the groups of sites you have studied and suggest reasons for both. Consider a range of factors which may affect bedload patterns within channels and channel efficiency – roughness (using your observed notes of each site), land use, geology.

Resources

'Measuring upland streams', *Geography Review*, Volume 11, Number 1, September 1997
'Rivers: vive la différance', *Geography Review*, Volume 10, Number 5, May 1997

Does river pollution increase downstream?

Starting Points ▶

1. How does water quality change downstream?
2. Is water quality associated with discharge?
3. How does land use and human activity influence water quality?
4. Do pollution levels decrease with distance from an urban area?
5. As pollution increases, does biodiversity decrease?

Geographical links to your syllabus

- Man's activity has a significant impact on water quality through agricultural practices, recreation and industry.

- The lower courses of rivers, flowing in wider, flatter and often more densely occupied valleys, tend to contain highest levels of pollutants.

- Rivers flowing through urban areas experience reductions in water quality.

5. Map detailed land use along the river up to 500m from each bank. Include type of agricultural land, urban land use such as waste ground, residential land, especially age of buildings, types of industrial use, recreation.
6. Locate field drains; any water abstraction points or water outflows into the river, sewage outflows etc.
7. Make an environmental quality assessment (see figures 44 and 53) along each bank.
8. Carry out a kick test to identify the aquatic species in the river at each site (see figure 15).

Primary Data Collection

1. Select at least 10 sites along a 5–10km stretch of river.
2. At each site measure width, depth, velocity.
3. At each site measure temperature, test for pH, dissolved oxygen content (see figure 13) and pollutants:
   hydrogen sulphide – use lead acetate paper (it turns black if lead is in the water);
   nitrates – use nitrate test strips;
   ammonia – turmeric paper turns yellow and pH values are greater than 7.0.
   There are various test kits available for other chemicals such as phosphates. Consult your science teachers.
4. Observe a qualitative index of pollution (see figure 14).

Figure 13 Biochemical oxygen demand

Collect a 500ml sample of water in two bottles. Measure dissolved oxygen in both bottles with a meter. Cover one bottle with aluminium foil for at least four hours, then measure dissolved oxygen in both bottles once again. The difference between the readings of each bottle is the BOD.

Figure 14 Qualitative index of river pollution

This simple observational test works well but remember that clear water can also be very polluted.

Clean		Polluted
1	**2**	**3**
clear water		very murky
stones clear and bare		stones covered with scum
no water weed		choked with weed
no grey sewage/algae		much grey sewage/algae
no scum/froth/oil		much scum/froth/oil
no rubbish		much rubbish
no sewage/strange colours		much sewage/strange colours

For each site assess the level of pollution. Add up the scores on a scale from 1 to 3 for each indicator to give a single figure.

⚠️

Things to look out for

Your personal safety is important. Treat polluted areas and materials with extreme caution. Always use rubber gloves.

You will find this topic easier if you select sites where there are contrasts in river pollution.

Think about sites upstream and downstream from a town or where there are distinct changes in rural land use. Survey your land use first to make a prudent choice of sites.

Secondary Data Collection

1. Find out if the local authority or Environmental Watch Scheme have surveyed the river. Use their data to consider any changes since the earlier survey.
2. Find out about any management plans for the river which will impact on water quality as well as discharge.

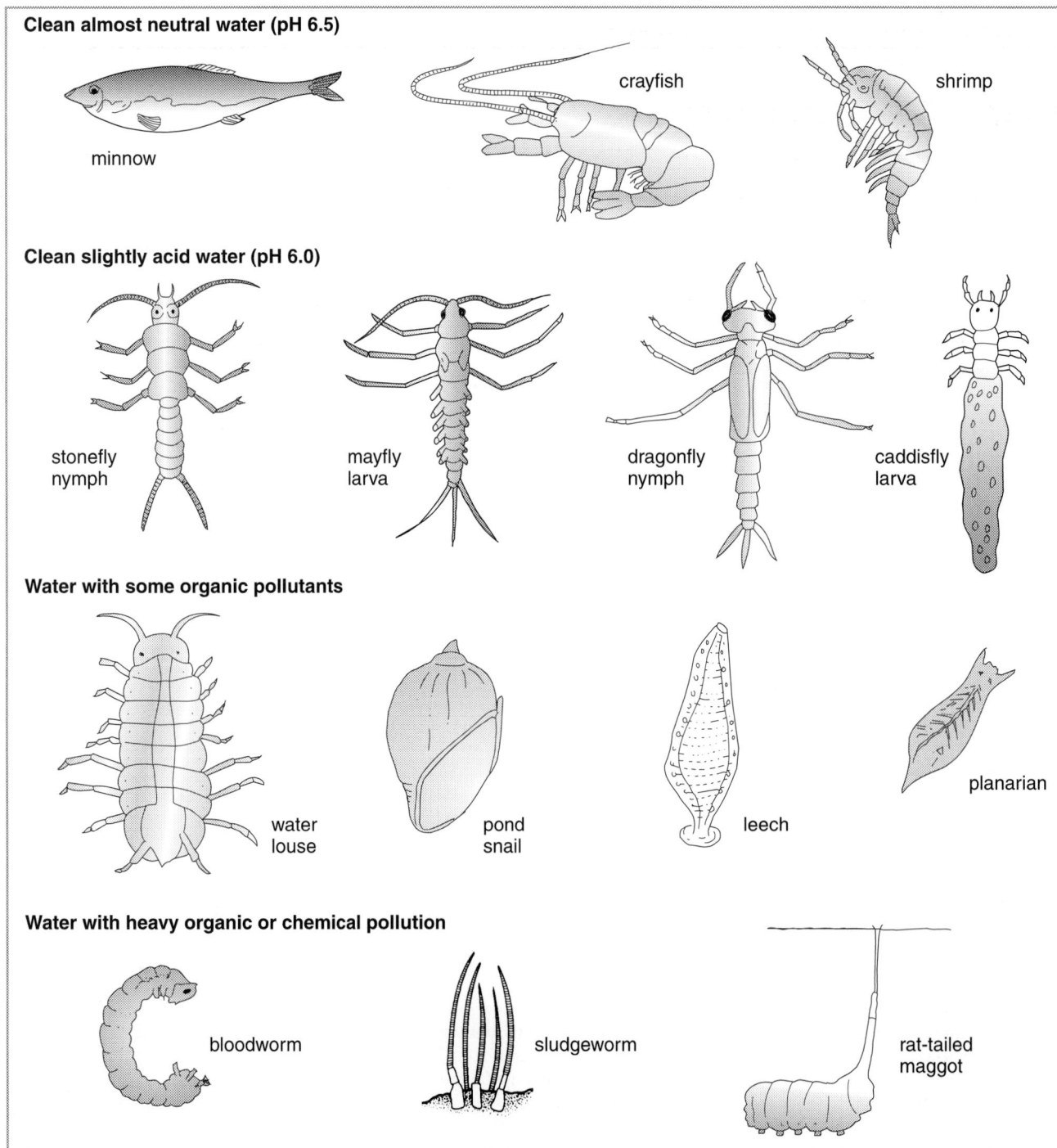

Clean almost neutral water (pH 6.5)

minnow   crayfish   shrimp

Clean slightly acid water (pH 6.0)

stonefly nymph   mayfly larva   dragonfly nymph   caddisfly larva

Water with some organic pollutants

water louse   pond snail   leech   planarian

Water with heavy organic or chemical pollution

bloodworm   sludgeworm   rat-tailed maggot

Figure 15 A key to some freshwater animals used as a pollution indicator species (not drawn to scale)

Ideas for recording and analysing data ▶▶

1. Draw and annotate a detailed land use map to show significant features relating to river pollution.
2. Calculate discharge at each site – average width × average depth × average velocity.
3. Graph relationships between variables such as those listed below and assess the correlation using Spearman Rank:

   discharge v BOD – the greater the discharge the higher the BOD

   BOD v velocity – the higher the velocity the higher the BOD

   BOD v temperature – the higher the temperature the lower the BOD.
4. Construct a series of line graphs one above the other to show changes in different pollutant concentrations downstream. (Sites downstream on the X axis.) Annotate to identify any links between one pollutant and another.
5. Graph the index from your environmental quality assessment against that for your pollution assessment for each site.
6. Draw a kite diagram to represent the number of each species identified in the kick tests at each site. Group the species to represent clean and polluted water. Use a Chi Squared test to find significant differences between species at particular sites. Link these results to your visual pollution index.

Interpretation and Conclusions

- How does pollution change downstream? What factors appear to have influenced those changes?
- What is the relationship between river bank land use and river pollution?
- How does water entering the river affect pollution levels?
- Do pollutant concentrations decrease with discharge?
- Is there a difference between pollution levels as the river flows through a town?
- How far downstream from an urban area are pollutants influencing water quality?

Resources

'Urban fieldwork locations', *Geography Review*, Volume 10, Number 5, May 1997

'Assessing river water quality', *Geography Review*, Volume 12, Number 3, January 1997

'Impact: water pollution monitoring', *Biological Sciences Review*, Volume 9, Number 1, 1996

Practical Ecology for Geography and Biology, by DD Gilbertson, M Kent and FB Pyatt, Hutchinson, 1985

Local authority environmental assessments

Environment Agency website: www.environment-agency.gov.uk/

National Rivers Authority website: www.open.gov.uk/nao/9495235

Atmospheric Research and Information Centre (ARIC) website: www.doc.mmu.ac.uk/aric

A comparison of the characteristics of streams of different orders

Stream ordering is one way of classifying streams within a drainage basin so that downstream change can be analysed.

Starting Points ▶

1. Does your drainage basin follow the theoretical relationships of stream order described in textbooks?
2. Do streams of the same order exhibit similar characteristics of bedload, channel efficiency, channel pattern?
3. How do channel characteristics change as stream order increases?

Geographical links to your syllabus

- As stream order increases drainage basin characteristics change – basin area increases; streams become longer; gradient decreases. How accurate are the theoretical relationships? Are those changes mirrored in changes within channels?

⚠ Things to look out for

Try to select sites just upstream from the confluence of streams. That way you will more closely link with drainage basin area.

Try to investigate a drainage basin of uniform geology.

Primary Data Collection

1. Within a drainage basin, select a variety of sites (at least five) on streams of different order. If you decide to compare say first and second order streams you can realistically study more sites on each. If you are working as a large group you can make a much more comprehensive study by coordinating results from more sites on different orders of stream.
2. At each site measure width, depth, velocity, wetted perimeter.
3. Measure gradient of stream at each site.
4. Sample 20 pebbles systematically across the stream. Measure a axis. Estimate roundness using Powers' index or Cailleux (see figure 9).

Secondary Data Collection

Use the OS 1:25000 or 1:50000 map to identify drainage basin details.
1. Measure basin area for streams in each order.
2. Measure length of streams in each order.
3. Record the changing geology within the drainage basin.

Ideas for recording and analysing data ▶▶

1. Map the drainage basins of all the streams and measure their area.
2. Draw channel cross sections for each site. Present cross sections of the same order together. Use the same scale.
3. Calculate width / depth ratio for each site. Calculate each hydraulic radius.
4. Graph width / depth ratios against stream order using semi-log graph paper. Repeat for relationships between stream order and velocity, discharge.
5. Calculate the roundness of pebbles at each site. Draw dispersion diagrams of roundness at each site. Find the mean and standard deviation so that you can compare results between different order streams.
6. Calculate bed roughness : average depth against average bedload size.

Interpretation and Conclusions

- Attempt to summarise the characteristics of streams in each order in terms of channel shape, width / depth ratio, gradient, velocity, discharge, drainage basin size, length of streams.
- How do these characteristics vary between streams of different order?
- Which variable appears to show the strongest link to stream order?
- Do you think that geology, land use or man's activity have affected the relationships you have investigated?

Resources

'Measuring upland streams', *Geographical Review*, Volume 11, Number 1, September 1997

An analysis of glacial and fluvioglacial deposits in a glaciated valley

Starting Points ▶

1. There are significant differences between size, roundness, angle of orientation and dip of deposits in different locations in a glaciated valley.

2. The origin of glacial deposits can be determined by their characteristics and location.

Geographical links to your syllabus

- A variety of fluvial and fluvioglacial deposits exist in glaciated valleys, from one or more ice advances.

- The characteristics of surface deposits can identify their origin and suggest the geomorphological history of the valley.

⚠ Things to look out for

Remember that your personal safety is the most important factor in fieldwork. Do not take risks in poor weather, on steep slopes, beside rivers.

Check the access to your sites. Do not trespass. Farmers are very supportive when they know what you are doing.

Preliminary Work

1. Research the glacial history of the area you are studying.
2. Draw a sketch map of the key glacial and fluvioglacial features which have been identified in the area.
3. If there are obvious examples of current processes at work in the area, such as scree, you may like to compare these with glacial deposits.

Primary Data Collection

1. Select accessible study sites on each of the geomorphological features you wish to investigate in detail. Consider feasibility of exposures in river banks, cuttings, quarries etc.
2. At each site sample 30 pebbles and measure angle of orientation and dip, a, b and c axes, roundness.
3. Collect a trowel sample of deposit (to half fill an ice cream carton) and label carefully. Dry the sample thoroughly then conduct a sieve analysis. Weigh the whole sample first then record the weight of each fraction after sieving.
4. Photograph each deposit carefully to show its location within the area.
5. Draw an annotated sketch of each site.

Secondary Data Collection

Many glaciated uplands have detailed geomorphological reports which outline landforms in terms of orientation and origin. These may support any conclusions you reach.

Ideas for recording and analysing data ▶▶

1. For each deposit you investigate draw a dispersion diagram of index of roundness. Identify median and interquartile range. Annotate the diagrams to highlight unexpected results.
2. Calculate mean and standard deviation of a axis of all pebbles at each site.
3. Fabric analysis. Draw a rose diagram (see figure 16) to show angle and orientation of dip of deposits (pebbles).
4. Draw a histogram of the results of the sieve analysis. Identify the distribution by its modal class (see figure 11). Use Chi Squared statistic to test for significant difference in range of particle size between different deposits.
5. Annotate photographs to suggest the origin and characteristics of the features you are investigating. Clearly identify their location on your sketch map of the area.

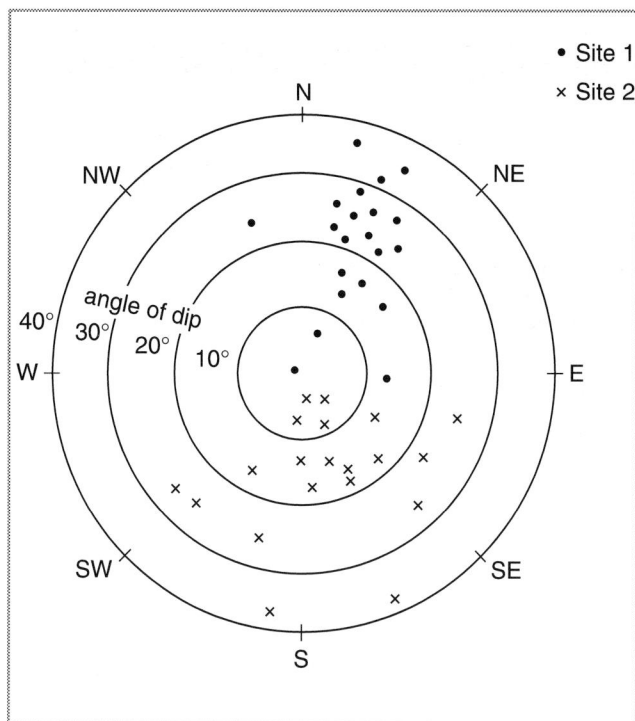

Figure 16a Rose diagram to show angle and orientation of dip of deposits

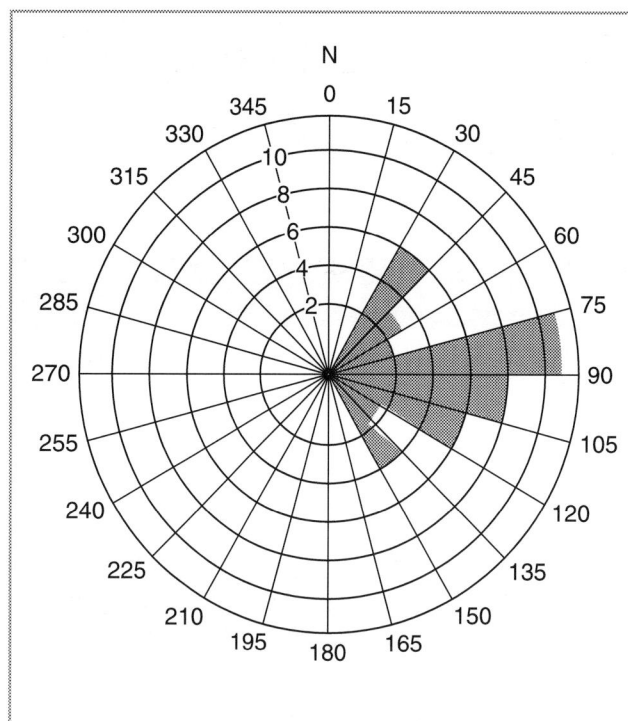

Figure 16b Concentric rings can also represent number of stones in every 15° of orientation

Interpretation and Conclusions

- Using your data, graphs and photographs describe and explain each feature you have investigated. Use the data to justify or confirm the type of deposit. To what extent is it typical of textbook features?
- If there are any unusual features, suggest explanations.
- How do glacial deposits vary? Compare their characteristics in detail. Does your evidence support or confirm the published research on the area?
- Compare and contrast current and historical processes acting on the landscape.
- Annotate a morphological map of the area to include the results of your investigations.
- Summarise your fieldwork with your own interpretation of the geomorphological history of the area.

Resources

Classic Landform Guides, published by the Geographical Association

Analysis of changes in vegetation across a sand dune system

Starting Points ▶

1. The amount and type of vegetation increases with distance from the sea.
2. Changes in vegetation are associated with the topography of the sand dunes.
3. Changes in vegetation are associated with increased soil depth, organic matter and moisture content.

Geographical links to your syllabus

- Sand dunes are typical examples of a primary succession (psammosere).

- Sand dune ecosystems usually exhibit a distinct and characteristic pattern of vegetation across a transect inland.

- As vegetation changes, soil characteristics also change.

Primary Data Collection

1. Establish a transect across a dune system from the sea inland. Follow a compass bearing to ensure that you work along the transect at right angles to the coast.
2. Survey the dune system using clinometer, tape and ranging poles.
3. The topography may influence your sampling decision. **Either** select your data collection sites by systematic sampling (ie at regular intervals).
   Or at each major change in slope / character of dune make a transect line at right angles to your line of investigation. Collect data at 0m, 5m and 10m along this line.

4. At each site use a quadrat to record percentage vegetation cover of each species. The ACFORN code (see figure 20) can be used if you are working on your own. Identify the type of species and note the number of different species. Alternatively use the point frame method to record vegetation (see figure 17).
5. Measure height of vegetation using sampling pins on a point frame.
6. At each sample site measure soil pH and soil depth.
7. Collect and label a soil sample at each site. Use this to test for humus content either with a sand smear test (see figure 19) or a more accurate laboratory method (see figure 18).
8. You may wish to extend your investigation by looking at microclimate across the dunes. In that case, record wind direction and wind speed at 0m and 1m above ground level. Several readings will be necessary to eliminate the impact of gustiness. You could also record soil temperature, air temperature at 0m and at 1m, and relative humidity. Remember that temperature in the shade is the most accurate. Direct sunshine will distort your results.

Secondary Data Collection

Many sand dune coasts are managed by conservation organisations. If that is the case in your area, search out the survey work – to help you particularly with plant identification.

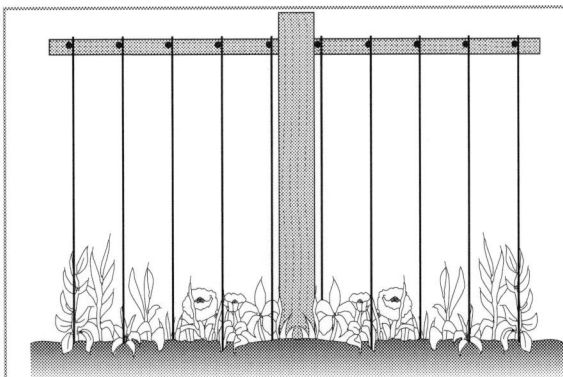

Ten pins are placed at equal intervals through a simple wooden frame. The species touched by each pin is recorded as a hit.

Figure 17 Point frame method of sampling

Figure 18 Soil moisture and organic content

Weigh a soil sample, dry it in an oven for 12 hours at 105°C then weigh again. This will give a measurement of **soil moisture**.

To find the weight of **organic matter**, place the dried sample in an oven once more, at 500°C for two hours.

$$\% \text{ organic matter} = \frac{\text{weight loss on burning}}{\text{weight of dry soil}} \times 100$$

⚠️ *Things to look out for*

Make your transect long enough. Many dune systems extend several hundred metres inland. Make sure you investigate the yellow and the older grey dune ridges.

Timing of your fieldwork is important – you will find it hard to identify plants between September and May.

Figure 19 Sand smear test for humus content in sand dunes

Take a palm-sized amount of sand / soil and dampen with water.
Allow to dry for two minutes then smear on to a patch of paper.
Allow to dry completely.
Rank the scale of results from
1 = very thin humus to
5 = very thick humus.

Figure 20 ACFORN code for vegetation abundance

Species cover
over 80% = A abundant
50–79% = C common
20–49% = F frequent
2–19% = O occasional
1% = R rare
0% = N none

Ideas for recording and analysing data ▶▶

1. Draw a detailed transect diagram to show the topography of the sand dune complex.
2. On the same page, underneath the transect diagram, represent change in percentage cover of vegetation. Shade the appropriate portion of a graph square to show percentage cover at each site. You could extend this to a kite diagram representing each species. Add details for height of vegetation at each site.
3. Represent soil depth and pH values along the transect. Also organic / humus content.
4. Draw scatter graphs to show the relationship between distance from the sea and organic content; pH; number of species; depth of soil; height of vegetation. If a correlation appears to exist, test this using Spearman Rank. Don't forget to test for significance.
5. Graph your microclimate data on another transect diagram. Only temperature should be represented by a line graph, so use a different method for relative humidity and wind speed.

Interpretation and Conclusions

- Does the type and amount of vegetation change inland?
- Is there a relationship between vegetation change and topography? Why should this occur?
- Do soil characteristics change with vegetation? Why?
- How does microclimate affect the changes in vegetation you observe?
- Can you identify any interrelationships between the different variables you have measured?
- To what extent are changes in your study area similar to those in textbooks? How are they different? Why?

Resources

Practical Ecology for Geography and Biology by DD Gilbertson, M Kent and FB Pyatt, Hutchinson, 1985
A key to plants common on sand dunes, Field Studies Council fold-out chart, Sally Edmundson, 1997; occasional publication 43

Assess the impact of human activity on the pattern of vegetation across a sand dune system

Starting Points ▶

1. The type and amount of vegetation becomes more developed with distance from the sea.
2. Vegetation development is affected by the location of car parks, picnic sites and footpaths.
3. As volume of human activity increases, the impact on vegetation increases.

Geographical links to your syllabus

- The ecosystem across a sand dune becomes more developed with distance from the sea.
- Leisure activities along coasts may have an impact on the development of vegetation.
- That impact has a distance–decay effect.

Primary Data Collection

1. Identify the focus or areas of intensive human use on a sand dune transect such as a car park. Sample at least 10 sites with increasing distance from that focus. Consider a systematic sampling technique. You could use a grid over a map of the whole area and sample at each grid intersection.
2. Record the site details in sketches or photographs.
3. Measure the width of paths leading from a car park at several sites with increasing distance from the car park. If paths appear to be worn down into gulleys you can measure depth of erosion. Place a taut tape across the path and measure depth of ground from the tape.
4. Measure compaction across the paths using a soil penetrometer. Alternatively, drop a spike from a known height and measure the penetration into the ground.
5. At each site make a vegetation transect across the path. Record species, percentage cover and height of vegetation.
6. Repeat this data collection at a point 5m from the path.
7. Count the number of pieces of litter in a 5m² area of footpath at each site.
8. Choose a fine weekend or Bank Holiday to count the number of people walking along the paths at each site. This should give you some idea of the greatest number of people using the area. Carefully consider how to overcome the problem of time of day.
9. Count cars in the car park at regular intervals during the day. Although you are bound to count some cars twice, this will still give you an indication of changes in intensity of use during the day.
10. Make an environmental quality assessment (see figure 21) at each site.
11. If there is a resident warden at the study area, find out about any management problems, issues and solutions.

Secondary Data Collection

1. In some nature reserves access is prohibited in order to protect vegetation. Try to acquire details of vegetation species and abundance from surveys conducted by the managing organisation involved. Remember to make your data collection compatible with any other data you collect.
2. Find details of the management plan for the dune system. Are the paths temporary or permanent? This may influence your site selection. Why are they established in particular areas?

⚠ *Things to look out for*

Your personal safety is important. Don't walk barefoot. Be aware of the tidal times and pattern.

If you are working in a managed system you may not be able to access areas unaffected by people. You will need permission to sample sites in protected areas.

Figure 21 Environmental quality assessment

Consider a variety of characteristics such as those below and allocate a score ranging from 5 for high quality environment to 1 for very poor quality of environment. Add your own ideas.
- noise
- litter
- trampling leading to damaged vegetation and erosion
- vandalism – broken notice boards and signs, broken fences
- intrusion of vehicles, bikes

Ideas for recording and analysing data ▶▶

1. Draw a detailed map of the area to show location of sites, car parks, picnic sites. Include path widths and details of areas with restricted access.
2. Draw a flow line map to show pedestrian flows along the path.
3. Draw cross sections of footpaths to show depth, pattern of vegetation species and percentage cover and also the compaction of paths. Use a modified kite diagram. Don't forget to annotate.
4. Map vegetation height across each transect. Draw several on one page, using the same scale so that you can compare. Beside each transect note the range, height and percentage cover of species of the undisturbed plots 5m from the path.
5. Graph amount of litter against distance from car park(s).
6. Draw a bar graph to show number of cars in the car park(s) at different times during the day.
7. Try a Chi Squared test to investigate differences between number of species near the car park and those furthest away, and on undisturbed and disturbed sites.

Investigation and Conclusions

- Does population pressure decrease from car park a) parallel to the sea and b) along the path towards the sea? Use evidence of litter, number of people, path cross sections.
- How much difference is there in vegetation along the path compared with the undisturbed sections?
- To what extent do the undisturbed sites follow the textbook pattern of vegetation in a sand dune system?
- Is there a relationship between volume of people and environmental impact?

Resources

'Environmental impact assessment', *Geography Review*, Volume 12, Number 2, November 1998
A key to plants common on sand dunes, Field Studies Council fold out-chart, Sally Edmundson, 1997; occasional publication 43
Environmentally Sensitive Areas website: iisd/iisd.ca/greenbud/sensuk.htm

How does beach material vary across a storm beach?

Starting Points ▶

1. The coarsest beach material is found at the back of the beach.
2. Beach material is sorted along as well as up the beach.

Geographical links to your syllabus

- Wave action on beaches results in the sorting of beach material.

- Constructive waves with strong swash build up storm beaches to considerable height above mean high tide.

- Strength of onshore winds and alignment of the beach influences the amount of sorting along a beach.

⚠ Things to look out for

Think carefully about your choice of storm beach. Are there several berms to analyse?

Is the beach wide enough to expect some movement of material along it?

Have you checked availability and type of tidal data?

Most important, does the tidal pattern make for a safe environment in which to work?

Primary Data Collection

1. To measure beach material, select three long transects from shore line to the back of the beach. One central transect plus one towards each end of the beach would be appropriate. Ensure the transects are at right angles to the sea by taking a compass bearing on a landmark at the back of the beach. Make sure that you always work along that bearing.
2. Choose your sampling technique carefully: i) a systematic sample every 20m which may or may not identify specific changes in slope; or ii) a subjective sample which selects sites to record changes associated with gradient up the beach. In that case sample beach material at the bottom and top of a berm. Remember that some storm beaches may have three or four ridges of increasing height. You will need at least 10 sampling sites along each transect.
3. Make an accurate survey of the slope of the beach along each transect line.
4. Place a quadrat at each sampling point to select 20–30 pebbles. Decide how to select each one, eg from the centre of each grid square or beneath each intersection. Measure a axis of each pebble.

Secondary Data Collection

1. Obtain a geology map of the area.
2. Gather information on direction and strength of prevailing winds, and details of tidal ranges; available from the Coastguard.

Figure 22 Beach slope and particle size

PARTICLE SIZE	BEACH GRADIENT
cobbles	24°
pebbles	17°
granules	11°
very coarse sand	9°
coarse sand	7°
medium sand	5°
fine sand	3°
very fine sand	1°

Source: Clowes and Comfort, *Process and Landform*

Ideas for recording and analysing data ▶▶

1. Draw a fully annotated map of the whole beach to show precise location of each transect.
2. Draw a detailed long profile of each transect. If they are presented on one page you can compare more easily.
3. For each transect draw dispersion diagrams using a axis of pebbles at each sampling site. Calculate mean, median and identify interquartile range.
4. Calculate the standard deviation for pebble size at each site. Ideally you should also calculate Standard Error since your sample size may be different from the much larger population of pebbles. However, you will need a lot more pebble samples to make this worthwhile. As it is, you are looking at differences between sites rather than analysing the population of pebbles in itself.
5. If you have taken a systematic sample for each site draw a scatter graph of mean a axis against distance up the beach. If there appears to be a relationship between size and distance, correlate using Spearman Rank, then test for significance. Treat each transect separately. Do they all follow the same pattern?

Interpretation and Conclusions

- Are the three beach profiles similar along the beach? What does this suggest about the wave processes operating along the beach?
- How do pebble sizes change up the beach? Does this fit with theoretical assumptions about wave energy in storm conditions?
- How do pebble sizes vary with gradient (see figure 22)?
- Is there evidence of sorting of material along the beach? In what direction? Why / why not?
- How does sorting (as measured by smaller interquartile range) compare at mean ordinary tide position and at higher levels up the beach? Would you expect more sorting at the ordinary tide position? Why?

Resources

If there are significant recent sea defences in the area, the local authority or Environment Agency may have data on wave strength and direction

Investigation of the factors influencing cliff erosion along a coast

Starting Points ▶

1. What processes are acting on the cliffs at different points?
2. What processes are acting on the beach below the cliffs?
3. How does geology affect cliff erosion?
4. To what extent is cliff erosion associated with human activity?

Geographical links to your syllabus

- A range of weathering, mass movement and hydrological processes simultaneously act in cliffs, although their relative importance changes over time.

- Differential erosion produces varied coastal landforms.

- Human activity is often instrumental in speeding up or slowing down rates of coastal erosion.

⚠ *Things to look out for*

Your personal safety comes first. DO NOT climb cliffs to investigate unusual features. You must ALWAYS work in a safe tidal area. Be aware of the speed of incoming tides, and do not allow yourself to become cut off by the tide. A pilot study is essential at high and low tides to plan your data collection in safety. Wear a crash helmet.

Select your sites thoughtfully depending on the emphasis of your study. If you are comparing different geology, choose at least three sites of each geological type. If you are looking at different aspect, choose half your sites facing in one direction and half in another. You must justify your choice of locations.

Primary Data Collection

1. Select at least 10 sites along your stretch of coastline. At 200m intervals observe a 20m section of cliff.
2. Draw annotated sketches of each section to identify all the mass movement processes acting on the cliff, eg mudflows, landslides, slumping, rock falls, dried gullies, rills, streams, undercutting, evidence of hydraulic action. Count the number of each in each 20m section of cliff. Look in detail – a quick glance is not enough.
3. If you are looking at geology, eg chalk, count the number of chalk particles found on the beach below the cliff. In till cliffs there may be specific material, eg granite pieces, which can also be counted on the beach below the cliff.
4. Below each cliff section, survey beach material. Place a quadrat at regular intervals, eg 10m from cliff base to shore line. Select 20 particles from each quadrat and record roundness. You could use Krumbein or Cailleux (see figure 9).
5. Measure the a axis of each particle you select.
6. If the beach material is too fine, collect a trowel sample, label carefully, and analyse in the laboratory. You need to air-dry each sample, mix or shake up the sample thoroughly, then weigh 100–150gms of each sample for further use. Keep the rest of your sample until you have finished all your analyses. Sieve your measured sample to identify different particle fractions (see figure 12).
7. Survey the beach below each cliff section to identify slope.

Secondary Data Collection

1. Obtain old maps to locate former position of coastline.
2. Obtain local geology maps.
3. Identify which organisation(s) is responsible for the stretch of coastline and find out about any footpath or management programmes affecting the cliffs and beach.
4. There may be several interested parties and a management plan for the area. Look for engineers' reports and coastal strategy reports. Organisations such as RSPB, National Trust, Environment Agency and the local planning authority may have some responsibility and therefore data or resources.

Ideas for recording and analysing data ▶▶

1. Draw a detailed map to show the location of all 20m cliff sections and significant coastal and terrestrial features in the area; also direction of maximum fetch, prevailing winds, important currents etc. Identify differences in geology if appropriate.
2. Draw annotated sketch diagrams, or take and annotate photographs to identify processes acting on each cliff section.
3. Tabulate the number of different processes operating on each cliff section. Draw a graph of processes in each section – several on one page – and annotate to identify similarities and differences.
4. Group the different processes into aerial and subaerial at each site. A Chi Squared test could identify whether or not there is a significant difference between types of process acting on different sections of the coastline.

5. Group processes further according to geology and analyse in a similar way.
6. Draw accurate beach profiles for the beach below each 20m section of cliff.
7. Add dispersion diagrams (see figure 4) beneath each profile to represent a axis of particles selected from the beach. Identify mean and interquartile range to analyse changes in size, both within each profile from sea to cliff face, and along the coast. Calculate standard deviation of pebble size below each cliff section.
8. Graph or tabulate pebble roundness from shore to cliff foot. Draw divided bars to show proportions of grain sizes from sieved material.
9. Are there differences up the beach? Along the coast? Is there a link between beach steepness and types of processes acting on the cliff?

Interpretation and Conclusions

- How do erosion processes change along the coast?
- Can you use all the evidence to comment on the rates of erosion along the coast?
- What impact does erosion have on a) the cliff and b) the beach material?
- Is there a link between rate of erosion and calibre of beach material?
- Does the size, roundness and sorting of beach material change from shore to cliff foot? Why? Are there changes along the coast? Why?
- How does geology affect processes, size of beach material, and rates of erosion along the coast?

Resources

Local authority strategy reports concerning coastal erosion
Newspaper reports of dramatic coastal erosion events

What are the interrelationships between vegetation, soil and microclimate on a shingle bar or a narrow sand dune system?

Starting Points ▶

1. How does exposure to the sea affect microclimate on the shingle bar?
2. Are there differences in soil characteristics on seaward and landward sides of the bar?
3. Are there differences in vegetation type, diversity and abundance on seaward and landward sites?
4. Why do these observed differences occur?

Geographical links to your syllabus

- Microclimates can vary significantly between exposed and sheltered coastal sites.

- The development of soils and increased diversity of species is expected where conditions are less exposed, less saline, less arid. Soil profiles should be more developed and vegetation more diverse.

⚠ *Things to look out for*

Local weather conditions, particularly wind, may not be typical of the study area. Remember that the climate conditions you record on one day cannot in themselves explain the existence or otherwise of soil or vegetation patterns. You are looking for *differences* between exposed and sheltered sites plus possible impacts of long-term climate patterns.

The speed of your data collection may be important to ensure that it is as comparable as possible. Mark out your sampling sites, then record all the microclimate data across the transect. Then go back to collect all the other soil and vegetation data.

Primary Data Collection

1. Establish a long transect from mean high tide over the bar to the same height on the sheltered side. Make a brief survey of the study area with annotated field sketches or photographs.
2. Use a systematic sampling method to locate at least 10 study sites at 20m intervals over the bar or sand dune system.
3. Take a small, clearly labelled soil sample from each site. Measure pH, moisture content, organic matter, and soil depth. You may have to dig a shallow hole with a trowel into sand or shingle to assess soil depth, or use a soil auger.
4. At each site record number and type of species, percentage cover, and height of vegetation.
5. At each site measure aspects of microclimate in the morning, midday and late afternoon; windspeed at 0m and 1m; temperature at 0m; soil temperature; relative humidity at 0m and 1m. This will give the pattern of differential climate change over the area and over time. Remember it's not the *actual* readings you take that matter but the *relative differences* between them.

Secondary Data Collection

1. Depending on your location there may be records of other surveys undertaken by the National Trust, English Nature, Countryside Council for Wales, or the Field Studies Council. These may be particularly useful if you are working when vegetation is not in flower and species recognition is difficult.
2. Obtain annual climate data for the local area.

Ideas for recording and analysing data ▶▶

1. Draw the simple surveyed section across the bar to identify sampled sites.
2. On or below the diagram, graph data for pH, percentage organic matter, soil moisture and soil depth.
3. Draw a kite diagram of vegetation species above the cross-section diagram. Represent percentage cover by shaded proportions of graph squares and small bars for vegetation height so that all your data is integrated. You will then be able to look for any interrelationships which appear to exist across the bar.
4. Annotate your diagram fully to highlight key links and changes across the system.
5. Draw scatter graphs to show the relationship between different variables and distance from the sea.
6. Consider whether any of your measured characteristics may be geographically linked with each other then draw scatter graphs to assess whether there may be any correlations to investigate. If a best fit line can be drawn, ie there is a *linear* relationship, consider a correlation test using Spearman Rank or Pearson's Product Moment correlation coefficient. (For the latter the data must be from a normally distributed population.) Remember to test for significance.
7. Look for links between your different sets of data, eg between percentage organic matter in soil and percentage vegetation cover; exposure as measured by windspeed and vegetation cover.

Interpretation and Conclusions

- Compare microclimate indicators near the sea with those on the opposite side of the bar or dune system. Are they significantly different? How? Why? Similarly compare soils and vegetation.
- For each of the microclimate, soil, and vegetation data sets, explain the pattern of results you have obtained. Make sure that you link the three aspects together. If you are to make reasoned statements, ensure that you have some statistical evidence to support your comments.
- Look for exceptions or unusual features in your data and suggest reasons for the results.
- Consider other factors which may affect your results, eg human activity, synoptic conditions at the time of your fieldwork, height of the shingle ridge or dune system, orientation of the coast in relation to prevailing winds.

Resources

A key to plants common on sand dunes, Field Studies Council fold-out chart, Sally Edmundson, 1997; occasional publication 43

What factors affect changes in soil characteristics down a slope?

Starting Points ▶

1. The appearance of the soil profile changes downslope.
2. Soil depth, soil moisture and soil acidity change downslope.
3. Land use influences the acidity and the organic and mineral content of soil.

Geographical links to your syllabus

- Soils are affected by mass movement processes downslope.

- Soils are affected by the processes acting within them, eg leaching, cheluviation, gleying.

- Temperature and rainfall patterns affect slope characteristics.

- Farming influences soil characteristics. Think about frequency of ploughing, application of fertiliser, pesticides, insecticides, intensity of farming methods, slope stabilisation and slope drainage.

Primary Data Collection

1. Select at least 10 sites down a slope. Make sure you justify your sampling technique.
2. Survey the slope carefully. Make annotated sketches of the study sites, looking particularly for features which may influence the soil, eg evidence of mass movement.
3. At each site record vegetation or land use, and slope angle.
4. Measure depth of soil using an auger. If you use a long thick wire such as a coat hanger insert the wire three to five times in a small area to avoid the possibility of hitting stones.
5. If you are able to dig soil pits discreetly you can record and sketch details of the soil profile – colour and depth of horizons, evidence of roots, fauna. Take soil samples from A and B horizons to test for acidity, nitrogen content, texture, and percentage organic matter. If you use a soil auger, carefully record changes in the upper and lower parts of the soil and test as above.
6. Use the FSC field method (see figure 24, page 33) to identify soil texture.
7. Measure infiltration rates at each site (see figure 23).
8. For soil moisture weigh a 50g sample, leave to dry in a low oven overnight, and weigh again. The difference in weight is the moisture content.
9. Interview the local farmer(s) / landowner(s) to establish how the land is used and what compounds may be added to the soil. Former land use is also important if it has changed in the fairly recent past.

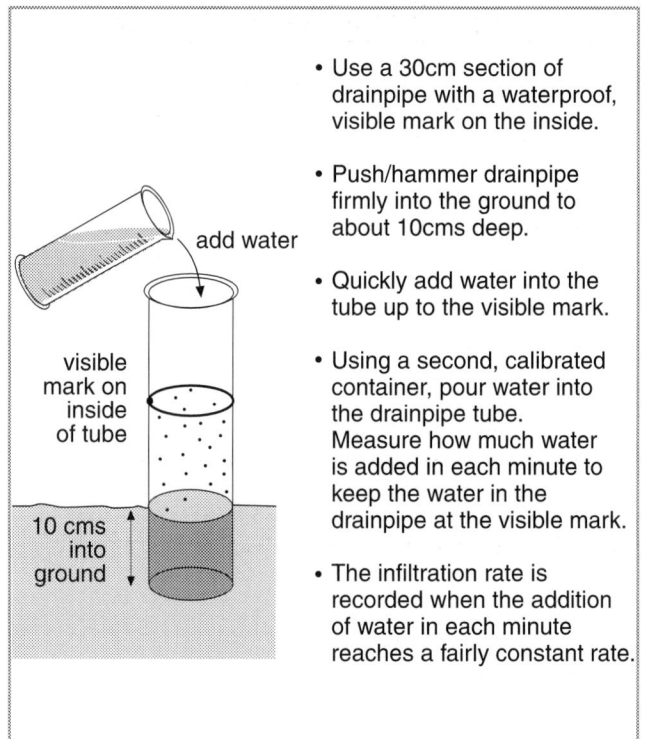

- Use a 30cm section of drainpipe with a waterproof, visible mark on the inside.

- Push/hammer drainpipe firmly into the ground to about 10cms deep.

- Quickly add water into the tube up to the visible mark.

- Using a second, calibrated container, pour water into the drainpipe tube. Measure how much water is added in each minute to keep the water in the drainpipe at the visible mark.

- The infiltration rate is recorded when the addition of water in each minute reaches a fairly constant rate.

add water

visible mark on inside of tube

10 cms into ground

Figure 23 Measuring infiltration rates

⚠️ *Things to look out for*

Length of slope. It must be long enough to show some changes, at least one kilometre in length and preferably with a substantial change in height.

Ideally the slope should have a fairly even gradient.

Don't forget the influence of geology and aspect.

Sampling. Systematic sampling will give you equal coverage over the slope, but you may wish to include particular land uses and slope angles. If so, consider a stratified sample.

Secondary Data Collection

1. Identify monthly temperature and rainfall patterns for the local area.
2. Find out about the geology of the area.
3. Find out about the geomorphological history of the area.
4. Identify land use history or obtain maps of former land use.

Ideas for recording and analysing data ▶▶

1. Graph the rainfall data for your local area.
2. Draw a long profile of your surveyed slope.
3. Map land use along the profile and identify area of human activity which may affect the soil characteristics.
4. Represent soil depth and details of soil horizons on the long profile diagram. Colour the horizons as accurately as you can.
5. On a second long profile, graph data for acidity, organic matter, texture, and soil moisture. Annotate the profile carefully to identify key elements of your data.
6. Draw scatter graphs to show the relationship between

soil characteristics and distance down the slope. If the graphs suggest a linear relationship, test using Spearman Rank or Pearson's Product Moment correlation coefficient. Remember to test your results for significance.

7. If you draw several graphs on one page you will be able to identify potential interrelationships between the variables you have measured. Consider graphing those interrelationships. *Beware of spurious correlations.* You must be able to support any links with geographical logic. Just because two variables change downslope does not necessarily mean they are connected as cause and effect.

Interpretation and Conclusions

- How do your measured variables change downslope? Consider anomalies as well as the overall pattern. Has antecedent rainfall had an impact?
- Are there links between variables in your particular area? Can you explain them?
- Does mass movement have an effect on soils down the slope?
- What impact does human activity appear to have on soil characteristics down the slope?
- How have farming practices affected soil characteristics? For instance, how significant are ploughing and fertiliser applications?

Resources

Climate data from local weather station or meteorological office
Practical fieldwork techniques, see website:
www.globe.org.uk/land/puplpage.html
Soil Association, 40–56 Victoria Street, Bristol BS1 6BY. Tel: (0117) 929 0661
'Underground Story: local variations in soils', *Geography Review*, Volume 10, Number 4, March 1997

Analysis of the development of vegetation along a hydrosere

Starting Points ▶

1. How does vegetation change from an aquatic to a terrestrial environment?
2. How do abiotic factors change as vegetation changes?
3. Does the area under study conform to the theoretical pattern descibed in textbooks?

Geographical links to your syllabus

- Primary succession is characterised by a series of stages.

- Vegetation at each stage is markedly different from other stages.

- Abiotic factors have an important influence on the development of vegetation through the succession.

⚠ *Things to look out for*

Your personal safety is important. Do not take risks in soft mud around the lakeside.

Be very careful not to disturb the vegetation. You will not be the first or the last to study the area, and excessive trampling disturbs what you want to investigate.

Primary Data Collection

1. Select a long transect from a freshwater lake inland, over at least 100m. Use a systematic sampling technique to identify at least 10 sampling sites along the transect.
2. At each sampling site, measure the pH of the soil rooting medium. Use the 'feeling test' (see figure 24) to identify soil texture.
3. Record soil depth at each site.
4. At each site record light levels at 0m and 0.5m above the ground.
5. Collect and label a soil sample from each site for laboratory investigation. Measure the soil moisture of each sample and the percentage organic matter (see figure 18, page 20).
6. Using a quadrat, record changes in vegetation along the hydrosere. Count the number of species; identify each species; note the percentage cover of each one.
7. At the shrub or woodland stage of the hydrosere, record species cover and variety in the herb layer. List each species and note their abundance in a $10m^2$ area.

Secondary Data Collection

If you are working in a conservation / SSSI / protected area, there may be results from earlier investigations of the hydrosere. If so, note particularly the location of different species at that time so that you can compare with the location of those species in your fieldwork.

Figure 24 Soil analysis using the 'feeling' test

Start by removing all the stones (>2mm) from your sample. This leaves the EARTH FRACTION. Moisten the soil. It is very important to thoroughly work the soil in your hands for a few minutes, perhaps adding more moisture, but not too much or it will be too runny!

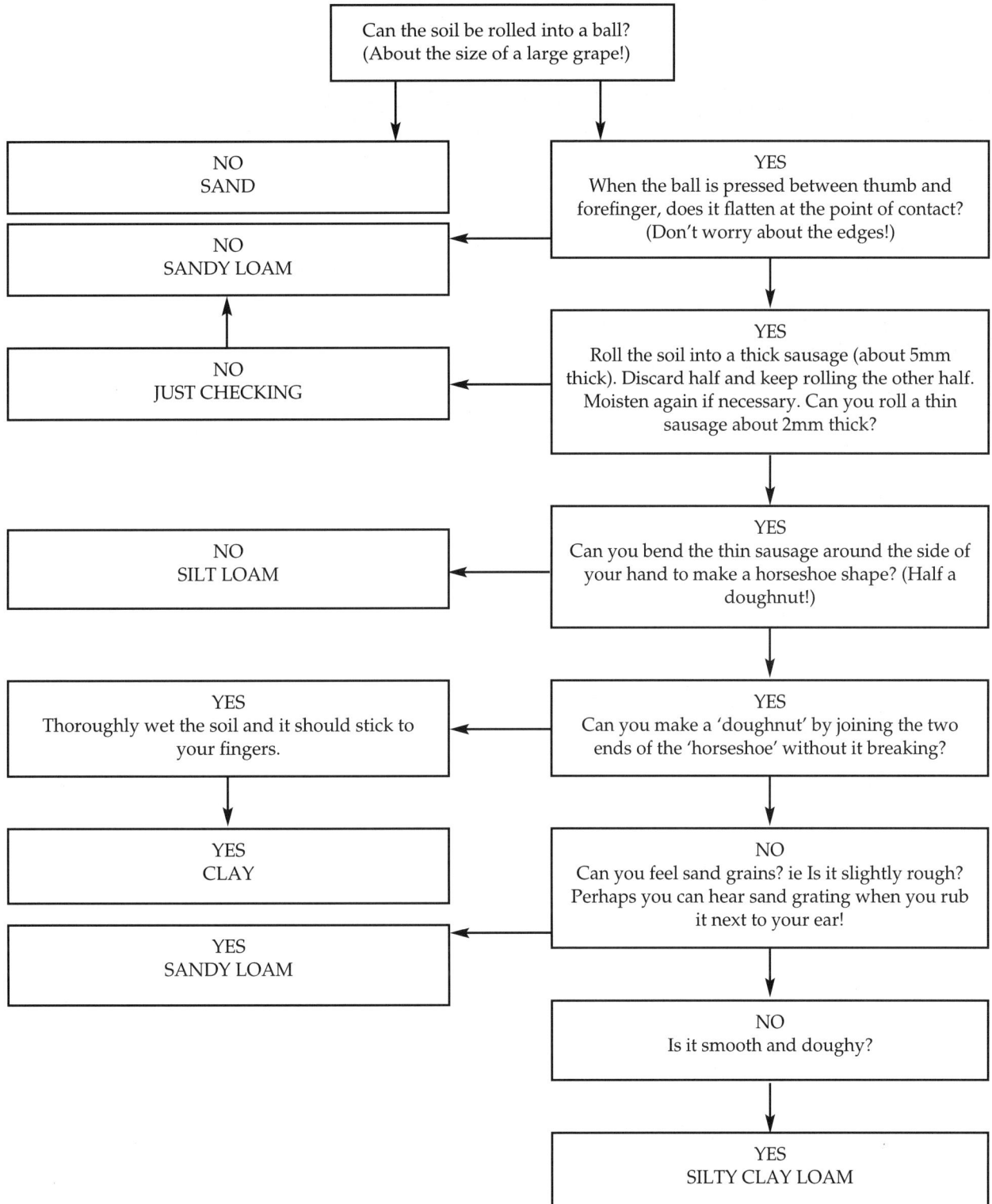

	Can the soil be rolled into a ball? (About the size of a large grape!)

NO SAND	**YES** When the ball is pressed between thumb and forefinger, does it flatten at the point of contact? (Don't worry about the edges!)
NO SANDY LOAM	
NO JUST CHECKING	**YES** Roll the soil into a thick sausage (about 5mm thick). Discard half and keep rolling the other half. Moisten again if necessary. Can you roll a thin sausage about 2mm thick?
NO SILT LOAM	**YES** Can you bend the thin sausage around the side of your hand to make a horseshoe shape? (Half a doughnut!)
YES Thoroughly wet the soil and it should stick to your fingers.	**YES** Can you make a 'doughnut' by joining the two ends of the 'horseshoe' without it breaking?
YES CLAY	**NO** Can you feel sand grains? ie Is it slightly rough? Perhaps you can hear sand grating when you rub it next to your ear!
YES SANDY LOAM	
	NO Is it smooth and doughy?
	YES SILTY CLAY LOAM

Ideas for recording and analysing data ▶▶

1. Draw kite diagrams to show the change in species and percentage cover over the length of the transect.
2. Calculate the species diversity at different sites along the transect using the Simpson-Yale index (see figure 26, page 36).
3. Beneath the kite diagram, graph changes in soil moisture, pH, soil depth and light levels. Annotate this transect diagram to identify links or interrelationships between the different variables you have measured.
4. Draw scatter graphs to show how different variables change with distance along the transect, eg changes in pH or soil moisture. If you think there appears to be a linear relationship use a Spearman Rank correlation to test your hypothesis. Don't forget to test your results for significance.
5. Consider the interrelationships between measured variables too, but make sure you can support them with geographical or biological reasoning.
6. Graph any secondary data at the same scale as your own observations to compare the development of vegetation over time as well as distance.

Interpretation and Conclusions

- How does vegetation change along the hydrosere? Does species diversity change?
- What are the links between biotic and abiotic variables?
- How has the vegetation developed over time? How old is the climax or plagioclimax vegetation?
- Is your location typical of those described in textbooks? Why, or why not?

Resources

Freshwater Investigations – a practical coursework guide by R Orton, S Haines and J Proctor, Field Studies Council occasional publication 36, 1995

Grasses Identification Chart, Field Studies Council occasional publication 33, 1993

'Biogeography: fieldwork techniques', *Geography Review*, Volume 10, Number 5, May 1997

How do woodlands influence microclimate? A comparison between coniferous and deciduous woodland

Starting Points ▶

1. Differences in light levels reflect differences in herb and ground layer flora from the edge of a wood to sites deeper into the wood.

2. There are significant differences in microclimate and associated vegetation beneath deciduous and coniferous woodlands.

3. There are significant differences in microclimate between the outside and inside of a woodland area.

Geographical links to your syllabus

- Vegetation can have a significant effect on the microclimate in which it grows.

- Species type and variety are associated with abiotic factors.

- There are links between soil processes and climate at a local as well as a regional scale.

⚠ Things to look out for

Choose an area that is large enough to have some potential for change from the edge to interior of the wood – at least one square kilometre.

You should ensure that you have suitable access to most parts of the woodland before you begin your detailed planning.

Whether or not the woodland is managed in some way will have an impact on your results. Man's activities should be carefully considered since there are few truly natural and undisturbed woodlands.

Primary Data Collection

Before you begin, decide which statistical techniques you wish to use. This will influence how many data collection sites you need, and the type of data required.

Choose your sampling method:

a) a transect from the edge to the inside of a wood
b) a selection of sites within the wood using a *systematic sampling method* – a grid over your detailed map will locate the sites.

The following are suggestions. You need not collect *all* the types of data.

1. Draw a detailed map of the woodland.
2. Microclimate data – measure light intensities at 0m and 1m above ground level; air temperatures at 0m and 1m; relative humidity; and wind speed.
3. Soil characteristics – soil acidity; soil moisture; colour; texture; and rooting depth. Record soil fauna you find. If you are investigating soils over a slope, survey the area for steepness and distance from valley bottom.
4. Vegetation – number and type of species; percentage cover of each species; height of shrub layer; and tree species at each site.

Secondary Data Collection

If your woodland is managed by a conservation group or Forest Enterprise, the wardens may have information on dates of planting, age of trees, and management techniques, such as tree thinning or coppicing.

Issues to consider in your data collection

- Temperature changes through the day and so does humidity. If you are looking at contrasts inside and outside the wood along a transect, you will find the greatest contrast in the middle of the day. Take all the temperature-based readings along the transect then go back to each site to complete the other data collection. Consider setting up thermometers in advance along your transect to make recording faster.
- Light readings will vary according to cloud cover as well as vegetation, so ideally you should work on either a cloudless sky, or a thoroughly overcast day.

- Wind speed is very variable according to gustiness. This will give only a general indicator of the shelter impact of vegetation.
- Soil moisture is related to the antecedent conditions in the area, but it is fair to assume that a small area would have experienced the same rainfall events.
- Consider the season – summer will give greatest contrasts, when most active growth of vegetation occurs. In winter, plants can be difficult to identify, and in spring, there may be very different flora visible compared with high summer.

Ideas for recording and analysing data ▶▶

1. Use detailed, annotated maps of the area to show distribution of vegetation or soil types.
2. Construct transect diagram to represent the relationships between all the data collected – light, wind, temperature, acidity, and details of vegetation. Annotate the main features on your graphs to summarise the data. Identify any unusual results.
3. Draw kite diagrams to show how vegetation changes with distance along a transect.
4. Use various bar, pie and divided bar charts. Draw several on one page so that you can make comparisons. Don't forget to draw attention to any results which are anomalous.

5. Use Spearman Rank correlation to test relationships, eg between light and air temperature.

   There are a number of methods you can elect to investigate species diversity in different environments.

   Mann-Whitney U test analyses the differences between data from different locations, in this case inside and outside the woodland.

   Chi Squared test is used to find out if there is a significant difference between the number of species at different points, either inside and outside the wood, or at different locations within the wood.

Figure 25   Jacard's similarity index

This measures the similarity of species between two sites.

$$J = \frac{c}{a + b - c}$$

a = number of spcies at site a
b = number of species at site b
c = number of species common to both sites

$$J = 0 \longrightarrow J = 1$$

no species in common   identical species in common

Figure 26   Simpson-Yale diversity index

This will compare species' variety in two different locations. The higher the value the greater the diversity.

$$D = \frac{N(N - 1)}{\Sigma n(n - 1)}$$

D = Simpson-Yale diversity index
N = total number of individuals
n = number of individuals per species

$$\text{General form of Simpson-Yale} = 1 - \frac{\Sigma(\text{percentage cover of each species})^2}{(\text{total percentage vegetation cover})}$$

Interpretation and Conclusions

- Attempt to explain the patterns of microclimate found beneath different vegetation types and densities.
- Are there differences in soil characteristics between vegetation types? Why?
- Can you explain the interrelationships you have identified in the graphs and diagrams. Can you explain the patterns shown on detailed soil / vegetation maps?
- Consider the limitations of your data collection and how that might affect your results.
- Suggest ways of taking your study forward or how your fieldwork could be followed up.

Resources

Institute of Hydrology: Maclean Building, Crowmarsh Gifford, Wallingford, *Grasses Identification Chart*, Field Studies Council occasional publication 33, 1993
A key to common plants in woodlands, Field Studies Council occasional publication 50
'Microclimate under trees', *Teaching Geography*, Volume 23, Number 4, October 1998
Practical fieldwork techniques, see website:
www.globe.org.uk/land/puplpage.html

How does microclimate in a valley differ from valley floor to summit?

Starting Points ▶

1. Upland areas are expected to experience higher wind speed, lower temperatures and lower relative humidity than valley bottoms.
2. Rates of change of temperature vary at different heights above valley floor.
3. Aspect has a significant impact on microclimate.
4. Changes in temperature throughout the day are greater on lowland sites than upland sites.
5. Land use may influence climate on hillsides.

Geographical links to your syllabus

- Mountain and valley locations experience markedly different climates.
- Such differences are associated with aspect, altitude, vegetation, wind speed and direction.
- Differences may be more pronounced in anticyclonic conditions.

⚠ Things to look out for

Choosing your study area – aspect can be influential.

Check with land owners regarding access to the hillsides.

Weather conditions – microclimate differences will be more easily identifiable in calmer conditions where there is less mixing of air, less wind and less turbulence.

You need sensitive, accurate equipment for microclimate work because changes can be very small.

Primary Data Collection

You will need to work in groups in order to collect data at similar times during the day. If you can, organise groups to work on opposite slopes of the valley. Never work on your own.

1. Select three sites on a slope with at least 100m height difference. Try to find open sites, or at least sites which are fairly comparable. Record the land use. Note the specific details of your actual measuring site, especially any features which may have an impact on your results.
2. At each site, at one-hour intervals, measure air temperature and relative humidity at 0m and 1m.
3. Record sunshine or cloudiness as you record temperature.
4. Measure wind speed more regularly, every 15 minutes, to give you an average figure. Remember that wind speed is very uneven because of gusts.

Secondary Data Collection

1. Collect recent climate statistics for the local area for the period of your data collection.
2. Note the synoptic situation during your fieldwork.
3. Obtain a large-scale map to show detailed contours of the slopes.

Ideas for recording and analysing data ▶▶

1. Draw a detailed cross section across the valley using a large-scale map and identify altitude and location of data sites.
2. Above the cross section, draw line graphs of temperature at each site for selected times in morning, midday and afternoon. Annotate the cross section to identify key features.
3. Draw bar graphs for relative humidity at each site for each time chosen.
4. For each site represent the changes in climate conditions through the day. Draw line graphs of temperature; average wind speeds per hour; and graphs for changes in relative humidity. Try to represent more than one site on a page so that you can identify different trends. Graphs do not have to be large to be effective.
5. Calculate standard deviation for temperature, wind and relative humidity for each site during the day.
6. Use the Mann-Whitney U test to compare daily changes in temperature from highest and lowest sites.

Interpretation and Conclusions

- You need to look at all your data, identify the differences, and suggest reasons for the patterns you find.
- How does temperature vary during the day at each site? Is there a link between wind speed and cloud cover or sunshine?
- How does relative humidity change? Why?
- Are there differences in daily wind and temperature profiles between each site? Why?
- Are there differences in temperature profiles at sites with different aspects?
- How does microclimate change from one side of a valley to the other?

Resources

Local weather data
Metfax: (0336) 400444 for the UK weather map

Analysis of the impact of weathering on gravestones

Starting Points ▶

Several factors influence the amount and rate of weathering of gravestones including age, geology, aspect, orientation and position of headstones, air quality as contrasted in urban / rural locations, local climate, vegetation within the graveyard, eg:

1. Is there a relationship between rate of weathering and type of geology of the headstone?

2. Is there a difference between rate of weathering and the microclimates of three contrasting graveyards?

Geographical links to your syllabus

- Weathering processes and rates of weathering depend on a variety of factors.

⚠ *Things to look out for*

Before you begin, make a courtesy call to the sexton of the graveyard to ask permission to work there.

Application of Rahn's index can be subjective so in the interests of consistency you should collect all the results yourself.

Primary Data Collection

1. The key to success here is not to have too many variables which affect your results. Make some key decisions about the focus of your investigation and try to make other influencing factors consistent. The table below summarises some of the options. A pilot study in the local graveyard is a must to help focus your ideas. Rahn's index (figure 27) is often used to classify degree of weathering, but you may find the student version (figure 28) easier to use.

2. It is essential that a substantial data set is collected. You cannot begin to reach conclusions without at least 100 graves recorded, preferably many more. If you are trying to keep the age of gravestones consistent, that will place significant restrictions on the availability of data. Remember too that some families renew their headstones following the death of the most recent family member. Ideally you need to make a stratified sample with an equal number of headstones in each category. In reality this will be difficult since many graveyards will be relatively limited once you have established your constant factors.

3. Decide whether you will accept tilting or flat headstones and justify your decision.

4. Use annotated photographs to illustrate the application of Rahn's index as a method of quantifying weathering.

Figure 27 Rahn's classification index

1. Unweathered
2. Slightly weathered – faint rounding of corners of letters
3. Moderately weathered – rough surface, letters legible
4. Badly weathered – letters difficult to read
5. Very badly weathered – letters almost indistinguishable
6. Extremely weathered – no letters left, scaling of surface rock

---

Figure 28 Students' classification index

1. Letters easy to read
2. Letters readable but slight weathering beginning around the edges
3. More obvious weathering over main part of gravestone
4. Small patches of lichen cover
5. Letters beginning to weather
6. More lichen found around edges
7. High proportion of letters weathered
8. Majority of letters weathered and large proportion of lichen coverage
9. Evidence of indents where letters were, but illegible
10. No lettering present

FOCUS OF INVESTIGATION	CONSTANT VARIABLES	FIELDWORK
1. Relationship between age of headstone and geology.	Locate on one site or look for similar sites in similar positions – urban / rural, similar air quality, similar distance away from roads, away from hanging vegetation.	Identify the different geologies to study. Record weathering on front of the headstone using Rahn's index. Note age of each stone.
2. Effect of air quality on rates of weathering. Urban / rural, near / far from a main road.	Choose same geology, eg only consider sandstones or slate, same age of headstones.	Record age of each headstone and weathering on each side using Rahn's Index. Use a quadrat to identify lichen species, percentage cover and variety on each headstone.
3. Effect of microclimate, eg E–W facing, impact of trees, height above ground.	Choose same geology, similar location, same age of headstones. Stones vertical or horizontal?	Record weathering using Rahn's Index at different heights from the ground, on E and W-facing sides. Look at headstones under trees, in more open situations, near bushes or shrubs. Measure relative humidity and temperature in the different locations, morning, midday and evening.

Figure 29 Tally chart for the investigation of weathering rates on different geology (numbers represent student's classification index)

	1700–1750	1751–1800	1801–1850	1851–1900	1901–1950	1951–2000
Sandstone	8 9 8 9	5 7 6 4 6 8	4 5 8	1 2 2 1 3 3 3 2 2	1 2 1 2 2 2 2 1 2 2	1 1 2 2 2 1 1 1 2 1
Marble	n/a	n/a	n/a	n/a	1 1 1 1 1 1 1 2 1	1 1 1 1 1 1 1 1 2 1 1 1
Granite	–	–	5 6 4	2 2 3 3 3 3 3 4	1 1 1 1 1 1 2 1 1	1 1 1 1 1 1 1 1 1 1
Slate	7 7 7 6 7 7	5 6 6 6 7	3 3 3 4	3 3 2 2 2 2 2	1 1 2 2 1 2 1 1 1 1	1 2 2 1 1 1 1 1

Ideas for recording and analysing data ▶▶

1. Draw detailed maps to locate precise locations of headstones. Categorise with a colour code to aid analysis.
2. Classify your headstones according to your hypotheses, then draw tally charts / graphs / diagrams of your data (see figure 29), eg degree of weathering *v* age; percentage lichen cover *v* age for different rock types; degree of weathering front and back of headstones *v* rock type.
3. If you have sufficient data you can compare groups statistically using Chi Squared, eg is there a significant difference between amount of weathering according to age groups, or according to surrounding vegetation, or between urban and rural graveyards?

Interpretation and Conclusions

- Does weathering increase with age? At what rate? Has that rate been consistent? Why or why not?
- How do rates of weathering on different geologies compare? Is there a significant difference? Why?
- Does weathering differ between urban and rural situations? Does location have an effect on rates of weathering?
- What is the impact of microclimate on weathering on different geology, and rates of weathering? What aspect of microclimate would seem to be the most influential? Unravel your data for range of temperature during the day, aspect, relative humidity, height from the ground, to assess the impact of each on weathering.

Resources

'Investigating urban weather', *Geographical Review*, Volume 5, Number 4, March 1992
Lichens: An illustrated guide to the British and Irish species by Frank S Dobson, Richmond Publishing, 1992
Lichens and Air Pollution, Field Studies Council occasional publication 34 British Lichen Society: c/o Botany Department, Natural History Museum, Cromwell Road, London SW7 5BD
Atmospheric Research and Information Centre (ARIC) website: www.doc.mmu.ac.uk/aric/
Portsmouth University research website: www.rgu.ac.uk/schools/mcrg/stgeo.htm
Most local authorities produce reports such as *A Review of Pollution in Cheshire*
Local air pollution data from DETR air pollution website: www.detr.gov.uk

Analysis of land use patterns in the central area of a town

Starting Points ▶

1. Land use changes from the Peak Land Value Intersection (PLVI) towards the edge of the CBD.
2. Land values decline with distance from the PLVI.
3. Building height decreases with distance from the PLVI.
4. Accessibility decreases with distance from the PLVI.
5. Distinct zones of land use can be identified within the CBD at ground level and on the higher floors of buildings.
6. Some shop types exhibit significant clustering compared with others.

Geographical links to your syllabus

- Bid-rent theory – the centre of the CBD is perceived as the most attractive location for business therefore commands the highest land values.

- The CBD has traditionally been the most accessible area of the city.

- Core-Frame model – the commercial city centre (inner core) surrounded by important but non-commercial land uses (outer core).

- The location of shop types depends on their response to competition, comparison shopping or convenience shopping.

⚠ *Things to look out for*

Remember that pedestrian density is highest at lunchtime. You either need lots of friends to help you, or think carefully where you begin on your own. If you start at the edge of the CBD in the morning, you could expect that any increases in pedestrian density would be reinforced as you get nearer the PLVI and as it gets later in the morning. The reverse will be true if you are working in the afternoon. Whatever you decide, you must be able to justify your method.

Primary Data Collection

1. Make a detailed map of land use in the centre of the town or city. Include the central area as well as the commercial or business area.
2. For each building record the number of floors, ie building height. If you want to make a more detailed study you could record and analyse the land use of different floor levels.
3. Use a systematic sampling method to select 30 shops from the whole area of the CBD. Make sure you include the shop likely to be at the Peak Land Value Intersection – PLVI. In many towns this is likely to be at or near Marks and Spencer. For each shop measure the frontage. (Simply pace the front of each shop and convert to metres later when you know how long your walking paces are.) Record the address of each shop. If this is difficult, note the shop name and look up the street number in the telephone directory.
4. Conduct a pedestrian count at various sites in the CBD. You could use the sites where you have selected the 30 shops to measure frontage. (See words of warning below.) A 10-minute period will be sufficient. Count in both directions, one after the other.
5. Choose three or four different types of shops, such as a newsagent, jeweller, shoe shop and estate agent. Map the location of every one accurately. This will form the basis of your nearest neighbour clustering investigation.

Secondary Data Collection

For each of the 30 buildings selected above, collect the rateable value or business rate from the Valuation Office Agency or Inland Revenue.

Ideas for recording and analysing data ▶▶

1. Group all the building uses you have recorded in the CBD into land use categories. Don't have too many groups. An 'Other' category is a useful catch-all for the odd ones which don't fit easily elsewhere as long as it does not become too large. Draw the land use map of your town or city using these categories. Don't forget that your map needs a scale. It may need to be quite small to fit on one page. It must be large enough to see without having to open up several sheets of paper within your folder.

2. Select 100m sections of the map from the PLVI to the edge of the CBD and calculate the proportion (percentage) of each land use category in each 100m section. Draw a composite bar graph of each of the sections to show how land use changes from PLVI to edge of CBD.
   Annotate the diagram to note any unusual features or confirm the expected pattern according to theory. Test for any difference between the most central section and the most peripheral section using Chi Squared. Don't forget to test for significance.

3. For all the 30 buildings selected, calculate rateable value per metre frontage. When you have confirmed the location of the PLVI, graph RV per metre frontage against distance from the PLVI.
   Annotate to identify pattern or unusual features. If the scatter graph indicates there may be some correlation between land value and distance, use Spearman Rank to test the relationship.

4. Graph building height v distance from the PLVI. Follow this up with a test for correlation, but beware there are often too many rank equal numbers to make this appropriate. Once again, don't forget to annotate your graph.

5. Construct a flow line map of pedestrian flows in the CBD. Note that the lines should be sufficently distinct to recognise different volumes of flow, but not so wide that they obliterate the street pattern.

6. On a separate map or maps, accurately locate the different shop types to investigate clustering. Measure the distance in metres between each shop and its nearest neighbour of the same type.
   Either i) Apply the nearest neighbour formula to each type of shop to give a measure of clustering.
   N.B. This formula depends on accurate calculation of area. You could simplify matters by calculating the average distance of nearest neighbour, then compare results between different shop types. While not strictly accurate this may be sufficient evidence to comment on distribution of shop types.
   OR ii) Calculate the standard deviation from the mean distance for each shop type. The lower the SD, the more clustering occurs.

7. Calculate the Index of Dispersion ID = Q / A where the range of values lies between 0 (highly clustered) to 1 (completely dispersed). See *Fieldwork Techniques and Projects in Geography*, Lenon and Cleves.

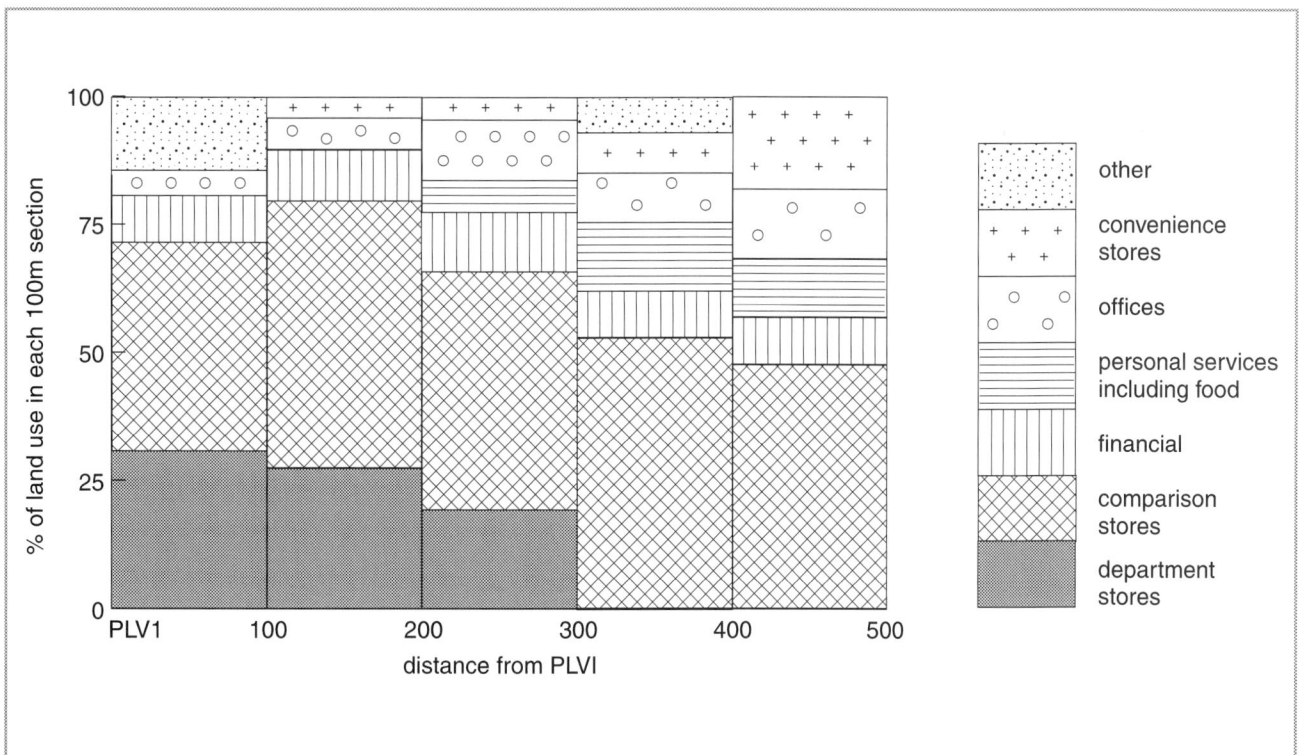

Figure 30 Composite bar graph to show how land use changes from the PLVI to the edge of the CBD

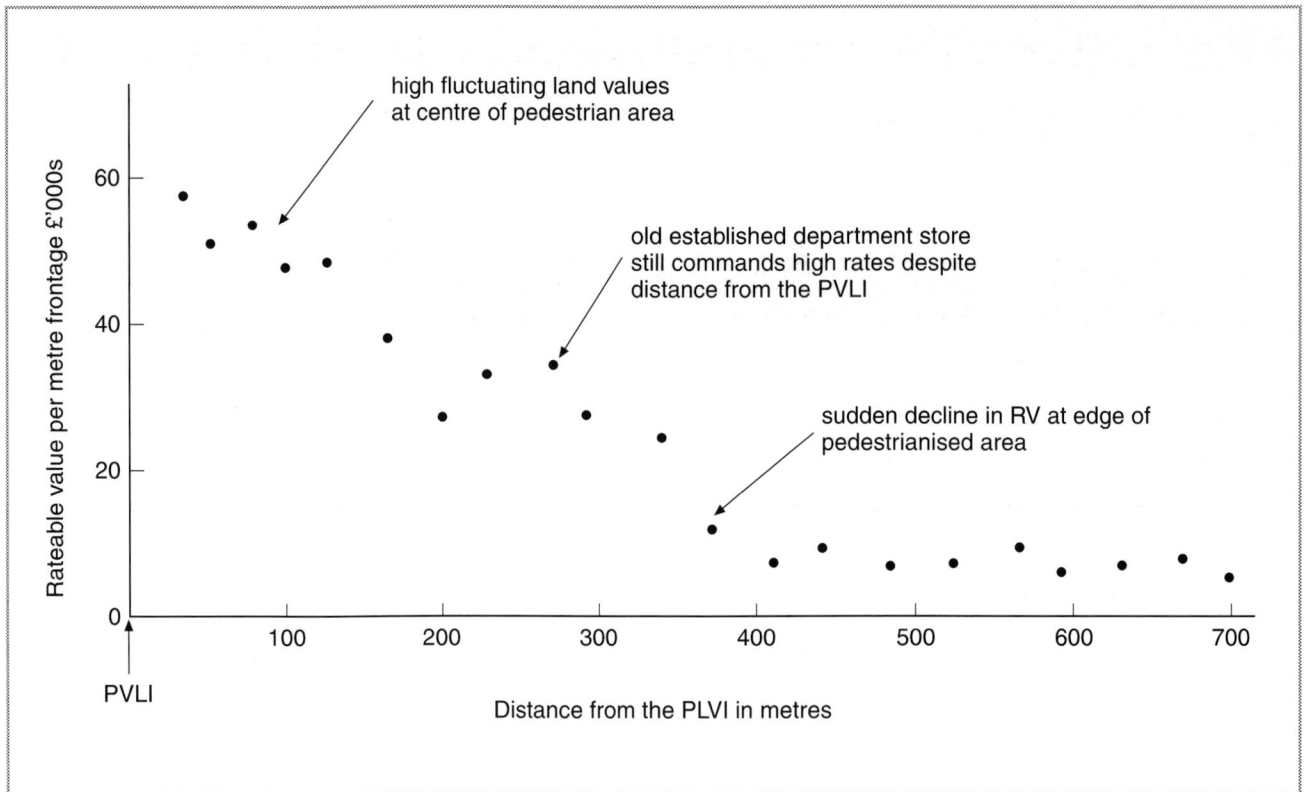

Figure 31 Scatter graph of rateable values per metre frontage and distance from the PLVI

Interpretation and Conclusions

Work through each hypothesis in turn and refer specifically to the appropriate graphs and maps. Comment on each one in relation to the theoretical patterns you would expect to find.

- How does land use change from the centre to the edge of the CBD? Is there an 'inner core' and 'outer core'? Were your results significant? If so, what factors influence the pattern in your town? Think about topography, rivers, communications, historical influences, civic functions or tourist functions. Can you explain the pattern you have identified?
- How do land values change across the CBD? Can your comments here be linked to those above? Are changes in land values mirrored by changes in pedestrian density? Why?
- Consider the change in building height. Are they significant? Why not? Where are new developments taking place? Are they high rise? What influences height? Historical or planning restrictions? Is this theoretical idea appropriate to British towns and cities? Why or why not?
- Do some shop types locate closer together than other types? Why? What is the evidence?
- How have local urban planning controls affected the land use patterns you observed?
- To what extent is your urban area typical of British towns? Suggest reasons for your view.

Resources

'Ecology of retailing', *Geography Review*, Volume 11, Number 3, January 1998
Goad map of town centre
Your local planning department may have pedestrian flow data
Local council offices may have ground floor plans and can give you total ground floor area
Town and county council website, eg www.cheshire.gov.uk and www.chester.gov.uk
Local structure plan

How applicable are urban land use models to a town near you?

Starting Points ▶

1. Distinct zones of different land uses can be identified.
2. Do land use patterns in your town show any similarity with theoretical land use models?
3. Land values decrease with distance from the CBD.
4. Building age decreases with distance from the CBD.

Geographical links to your syllabus

- Urban land use models summarise patterns in towns.
- Bid-rent theory.
- Urban areas change through time.
- Land values are associated with levels of accessibility.

⚠ Things to look out for

Remember, your personal safety in towns is most important.

Beware. Theoretical land use models were developed with reference to cities, in particular, American cities (Burgess, Hoyt). If you choose to use these, justify your choice carefully. Mann's model was based on industrial cities in Britain. Look at more recent models which seek to explain current urban patterns, eg core-frame concept.

Remember that planning regulations are much more strict in the UK than in American cities.

Be realistic – don't choose a town which is too small. Consider the context in which the urban model was developed in the first place. If you are investigating an urban area where factors other than industry have been significant, you may not necessarily refer to a model. Think about land use patterns and generate your own model which could be applied to towns or cities with similar backgrounds.

Make sure you can get to different areas safely.

Primary Data Collection

1. Identify the central point in the town. Choose four routes in different directions from town centre to edge of urban area to form four transects.
2. Along each route select 200m sections in which to record land use of each building. The number of sections depends on the size of the town, but you need at least four along each transect.
3. In each of these sections select five buildings – one every 50m – and note address, age of building and width of frontage.
4. Observe and map broad land use patterns across whole urban area. Identify different types of residential areas. This needs to be a generalised map. Don't get too sidetracked with small-scale changes, such as the occasional factory unit located in an unusual place. You should make a note of such land uses, and use the information in your discussion.

Secondary Data Collection

1. Collect rateable values or bands or business rates for each building you have recorded above. You will need the addresses to select the correct building from the lists supplied by the council or Inland Revenue.
2. Use 1:25000 OS map to support observations of land use over the whole area.
3. Town councils often have land use maps available in the Planning Department.

Ideas for recording and analysing data ▶▶

1. For each route or transect draw detailed land use sections. Use a sufficiently small scale to represent all the sections for each transect on one page.
2. Calculate rateable value per metre frontage for all the buildings selected. This will give an indication of land value, although the actual rateable values are based on floor area, not width of building.
   Draw a line graph of RV per metre frontage *v* distance from the centre of town.
3. Use Spearman Rank to test correlation between rateable values and distance from town centre.
4. For each route calculate the percentage of different land uses in each 200m section. Draw a compound bar chart for each route out of the town (see figure 32).
5. Is there a significant difference between land use percentages in the first 200m section compared to the last 200m section at the edge of the urban area?
6. Construct a land use map of the whole urban area using your detailed transect information and the general observations you have made.

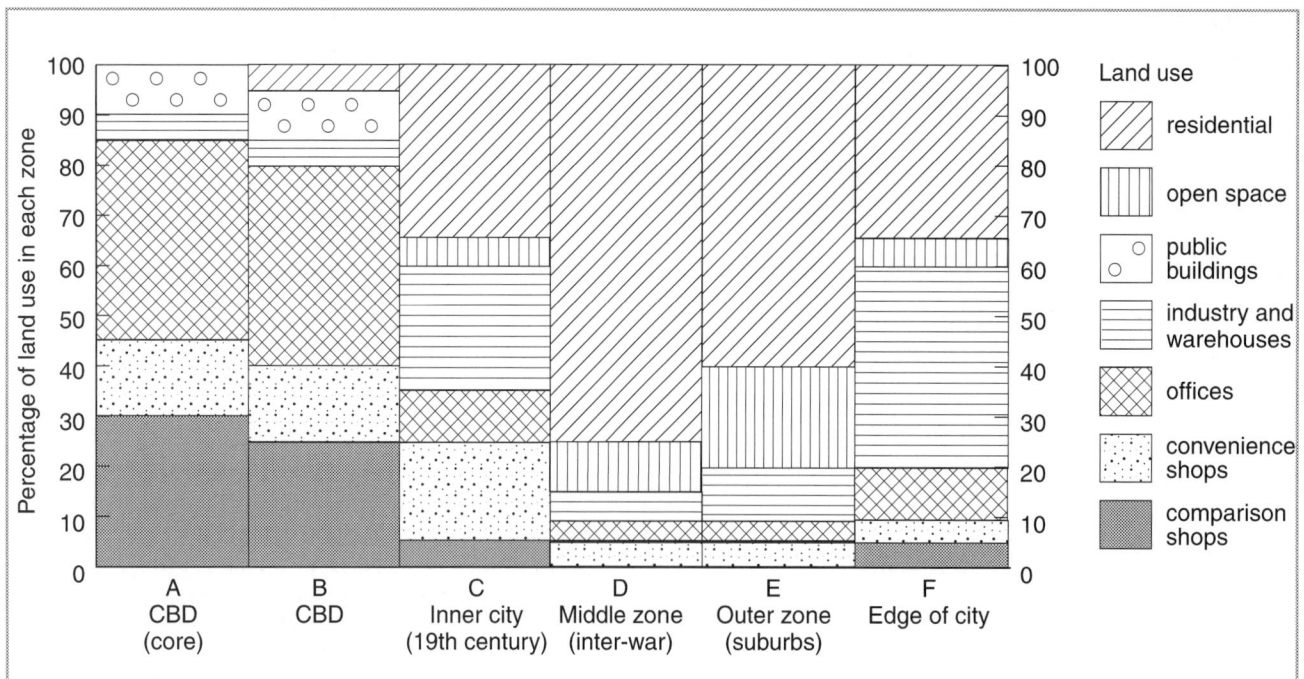

Figure 32 Idealised pattern of land use in a British city

Interpretation and Conclusions

- Are there clearly identifiable zones of land use in your town? Why have they developed? (Be careful – this is not your History coursework.)
- Are there changes in land values from the centre to the suburbs? Why?
- What similarities and differences are there between your land use pattern and that of the theoretical model you have selected? Can you summarise and comment on why there are similarities? Where and what are the differences? What processes have led to these differences? Could this be the source of further research?
- What other factors have influenced the land use pattern of your town? Consider topography, communications and historical influences.

Resources

1:18 182 OS street atlas will give large-scale detail about land use in selected areas. Local street atlases are available at the 1:16 313 scale
Goad maps of town centre
City / town planning departments have details of land use in their areas
Local structure plan
Try searching the town website(s)

An exercise to delimit the central area of a town

Starting Points ▶

In most towns there is a distinct change in character between the commercial centre and the area around it – the 'transition zone'. This exercise aims to identify as precisely as possible where the CBD ends and the 'transition zone' begins.

1. There are differences in land use and building occupancy between the CBD and the transition zone in a town or city.

2. There are differences in accessibility between the CBD and the transition zone.

3. Land values and environmental quality differ significantly between CBD and transition zone.

4. Building height changes from the CBD to the transition zone.

5. There is often more open space in the transition zone than in the CBD.

Geographical links to your syllabus

- Land use models of cities describe significant changes between CBD and transition zones – Burgess, Hoyt, Harris and Ullman, Mann, the core-frame concept.

- Theories of bid-rent reflect differences in accessibility between CBD and transition zone.

- Modernisation of city centres often includes pedestrianisation and changing traffic flows. This has an impact on the character of the transition zone.

- While the CBD remains the busiest area with highest commercial land values, its accessibility is restricted to the outer limits. Transport facilities are often concentrated in the transition zone as near the CBD as possible.

⚠ Things to look out for

A preliminary visit is crucial to make general observations about land-use changes in and around the central area or CBD of a town.

To be able to delimit the CBD you must ensure that your transects cover a good section of the transition zone as well as the CBD.

Remember that urban land use models generally applied to American industrial cities and may not be wholly relevant in Britain. It is unlikely that the precise patterns will be the same, but the underlying processes may apply.

Primary Data Collection

1. From your preliminary observations, identify four or five transects, about 500–1000m in length, that appear to include aspects of the CBD and the transition zone. Select these transects from different directions around the town centre. Make sure that each one starts at the same distance from the Peak Land Value Intersection or centre of the CBD.

2. Along each transect, record building height, building use (including vacant properties) and open space.

3. Select 20 buildings at regular intervals through each transect and measure the frontage. Record their precise address. The number in the street is important, but you could record name of company/business then check the address in a telephone directory.

4. Make an environmental problem assessment (see figures 33 and 34) at regular intervals along every transect, eg every 10–20m. The more sites you have the better.

5. Using a large-scale map, record one-way systems, car parks, car parking costs and restrictions in the area of your transects.

6. Map the amount and type of open space in the area around each transect.

Secondary Data Collection

1. For every building you have measured, collect rateable values or business rates from council or Inland Revenue.

2. Collect information on any new developments in your town, particularly those affecting the area of your transects.

Ideas for recording and analysing data ▶▶

1. Draw land use transects using the different land use categories you have identified. Arrange them carefully on the page so that you can compare them easily.
2. Calculate land value per metre frontage for the 20 buildings selected from each transect. Draw line graphs of land value per metre frontage v distance from the CBD end of your transect. If you can draw a best fit line and there appears to be a linear relationship, think about a correlation technique such as Spearman Rank.
3. Graph building height along each transect. Graph the number of environmental problems against distance from the beginning of each transect. Incorporate as much as possible of your transect details on one page so that you can identify interrelationships between the different characteristics.
4. Map the transport network, one-way systems, parking costs and restrictions.
5. Classify and map areas of open space in the areas around each transect.
6. Integrate all your recorded data and your personal judgement to create a map of the urban area to show the primary and secondary cores of the CBD and the transition zone (see figure 35).

Interpretation and Conclusions

- Attempt to delimit the CBD on each transect using each variable in turn – land value, environmental quality, type of land use, building height. Can a particular point / narrow zone be consistently identified along the transect? Justify your location of the edge of the CBD. Are there any observations which do not support your location for the limit? Why?
- Do all the transects follow the same pattern? Why or why not? Do any physical features affect the pattern of land use?
- Which of the variables observed do not fit the consistent pattern?
- Use observations of parking costs, location of car parks, traffic routes, one-way systems to support your conclusions about the limit of the CBD. New transport facilities are built as near as possible to the CBD but are often part of redevelopment or modernisation schemes in around the CBD or transition zone.
- Remember to relate your explanations to historical growth and theoretical ideas of land value.

Figure 33 Environmental problem assessment

Make a detailed assessment of the number of environmental problems at each of your recording sites.

One point is attributed for each of the following problems:
- illegal fly postering (each distinct item)
- litter (five major items)
- graffitti (each distinct item)
- broken window (each pane of glass)
- broken / protruding pavement (each distinct item)

For each site record the total score ie the number of environmental problems present at each site.

Resources

Details of transport network in the town
Planning proposals – the structure plan for the town or city from the local planning office
Rateable values or business rates from Inland Revenue

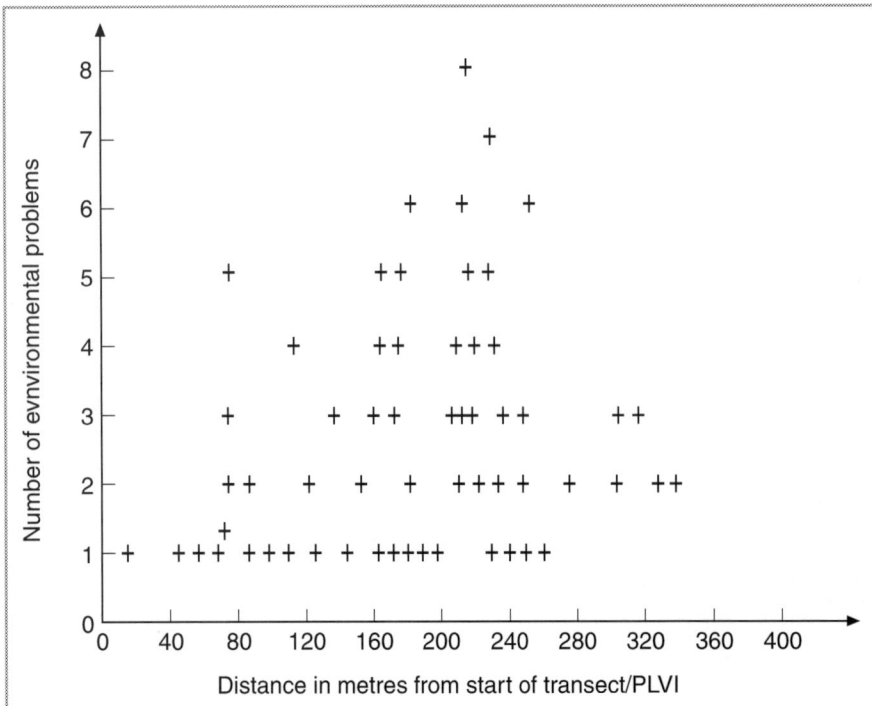

Figure 34 Scatter graph showing the number of environmental problems and distance from the PLVI

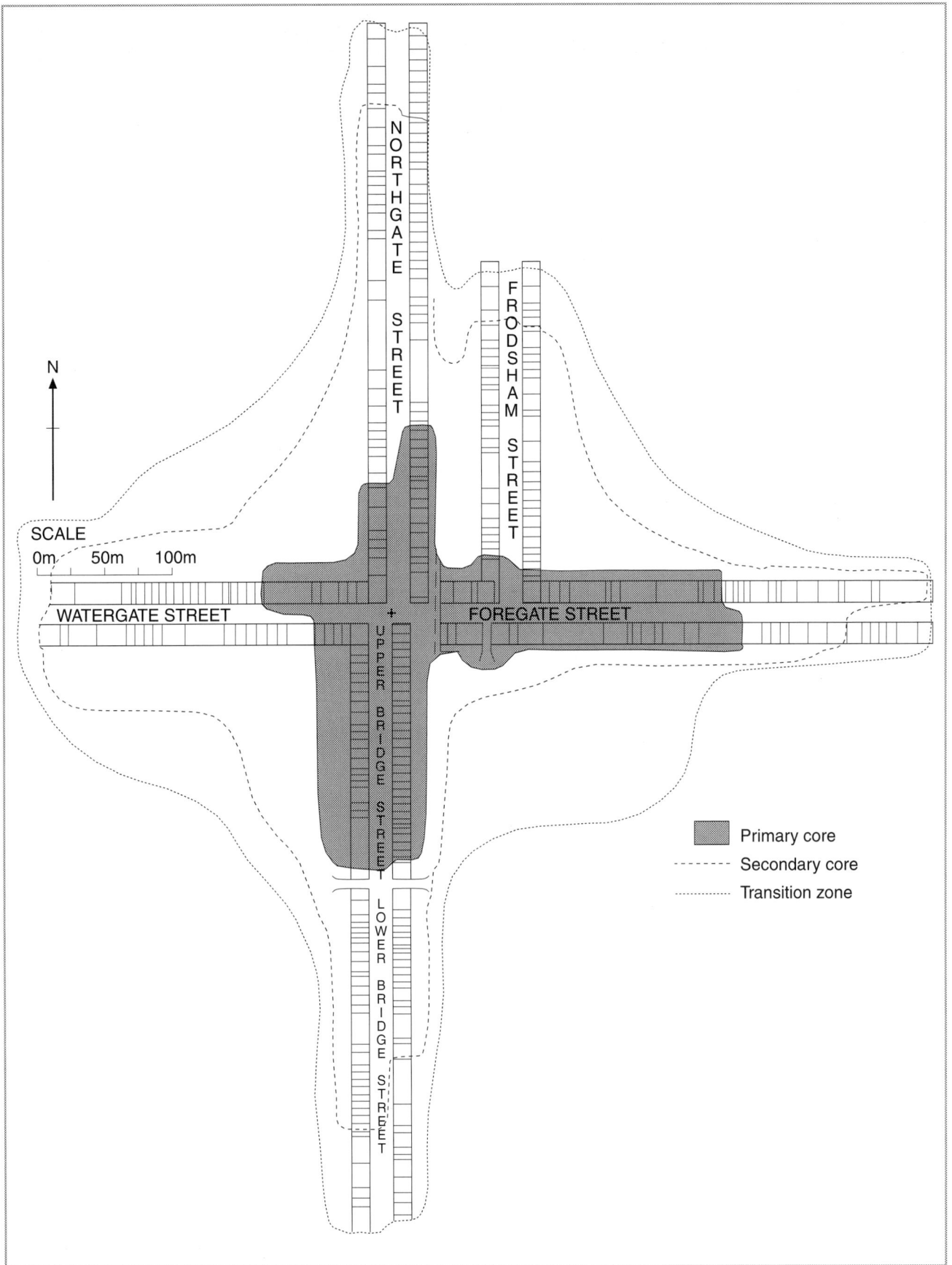

Figure 35 Map showing the extent of the primary and secondary cores and the transition zone in Chester

Analysis of redevelopment and land use changes in a town centre

Starting Points ▶

1. The amount of change increases with distance from the Peak Land Value Intersection.

2. Most changes in land use have taken place at the edge of the CBD.

3. Most changes in land use have been from residential and industrial uses to commercial and transport uses.

4. Changes in land use are often restricted to particular parts of the city.

5. Different planning strategies have led to different types of change at different times.

6. Increased numbers of vehicles have led to major structural changes in the central areas of towns and cities.

7. Changes in building function, landscape and infrastructure have increased the sphere of influence of the town or city.

Geographical links to your syllabus

- Space in the CBD has been under increasing pressure as our standard of living has risen and we spend more on consumer goods. Shops wished to expand in size and number. This was typical of the 1970s and 1980s.

- In the 1990s, changes in shopping styles led to a decline in some city centres and to expansion in out-of-town centres of various sizes.

- Post-war redevelopment of poor-quality housing and industrial sites in the inner city resulted in changes in land uses.

- There has been a decline in factories and industrial units, originally located around the central areas in the transition zone, particularly in the 1960s and 1970s.

- Demands for greater access to the city centre by public and private transport led planners to seek space for car parks and bus stations in the 1970s and 1980s, and easy access from park-and-ride schemes in the 1990s.

- Congestion in city centres led to building of inner ring roads – as close to the centres as possible – typical in the 1970s.

- This congestion also led to pedestrianisation and modernisation of the CBD.

Primary Data Collection

1. Place a grid (eg of 1cm squares) over your large-scale street map to identify 50 points or locations to be analysed. Note the grid reference of each point.

2. At each point observe the current land use and age of buildings. Classify into groups, eg shops, offices, finance, civic buildings including hospitals, town hall and library, industry, transport, residential (this could be subdivided depending on your observations), open space, derelict land etc.

3. You need to make general observations of land use across the whole town to put your detailed changes into context. Using a street map cover as much of the area as possible and record land use. Do not try to incorporate single buildings which are untypical of an area or street.

4. Attempt to make an environmental quality assessment for each site (see figure 44, page 58) and support your assessment through photographs.

5. Identify the main transport routes today and 30 years ago.

6. Conduct a questionnaire of 50 local people who have lived in the town since the 1960s. Is the area better, worse or the same in terms of noise, air pollution, shopping, quality of buildings, leisure facilities and ease of travel?

7. Can you make a subjective assessment of the impact of land use changes on the local people, economy and environment? How may each have been affected by any of the observed changes in land use?

⚠️ *Things to look out for*

The size of your grid is important. Too small and you have too many intersections and too many points. Too large and you have insufficient points to make significant analysis.

Check the availability of maps from the 1950s and 1970s so that you can precisely identify which dates you will focus on.

Make sure that you also have the appropriate Kelly's Directories. Most central libraries will have them, but they may not be readily available on the shelves. The last Kelly's Street Directory was published in 1974. Current Kelly's Directories list businesses in the local area.

Secondary Data Collection

1. Refer to old street plans of your town. There will be a map from the 1950s and probably the 1970s. Use the grid references for each of your 50 points, and the classification you have created, to identify what the land use was at the date of the maps.
2. Refer to local planning department records and a Kelly's Directory for the periods you are studying. Collect details about the nature of the land use, ie what type of shops were there, what kind of factory or office, what type of housing.
3. Local newspapers of the period may give details or photographs of some of the areas where there has been change so that you can collect more details.
4. Find local authority surveys of pedestrian density and traffic flows at different times in the recent past.
5. Census data will provide details of the employment structure of the town or city.

Ideas for recording and analysing data ▶▶

1. For each location identify the changes at the dates you have been investigating (see figure 36). Use the classification you created.

Figure 36 Table of land use change

SITE NUMBER/REF	LAND USE IN 1956	LAND USE IN 1976	LAND USE IN 2000
1	shop	shop	shop
2	shop	finance	office
3	shop	civic	office
4	civic	civic	civic
5	office	library	library
6	factory	transport (road)	transport (road)
7	factory	leisure	leisure
8	residential	derelict	transport
9	transport	derelict	civic
10	civic (hospital)	transport (car park)	transport (car park)
11	residential	residential / redevelopment	residential / second redevelopment

2. Construct a matrix to record changes from one year to the next (see figure 37).

Figure 37 Matrix of land use change

	LAND USES IN 2000								
LAND USES IN 1976	*Shops*	*Offices*	*Finance*	*Civic*	*Industry*	*Transport*	*Residential*	*Open space*	*Derelict*
Shops	12	2	1	0	0	0	0	0	2
Offices	1	8	2	0	0	1	0	1	0
Finance	3	2	6	0	0	0	0	0	0
Civic	0	2	0	6	0	1	0	0	0
Industry	0	0	0	0	3	8	5	7	3
Transport	0	2	7	1	0	9	0	1	0
Residential	0	1	0	0	2	8	14	7	1
Open space	2	1	0	0	1	6	3	5	0
Derelict	2	4	1	1	3	8	2	4	3

3. Use Chi Squared to investigate whether there have been any significant changes in a particular area and/or between particular types of land uses (see figure 38).
   Is there a significant difference between the amount of land in these categories in 1974 and that in 2000?

Figure 38 Application of the Chi Squared statistic to test whether there is a signicant difference in land use between 1976 and 2000

	1976 OBSERVED	EXPECTED	2000 OBSERVED	EXPECTED	TOTAL ROW
Shops	17	18.5*	20	18.5	37
Offices	13	22.5	22	22.5	45
Financial	11	14.0	17	14.0	28
Civic	9	8.5	8	8.5	17
Industry	26	17.5	9	17.5	35
Transport	20	30.5	41	30.5	61
Residential	33	28.5	24	28.5	57
Open Space	18	20.5	25	20.5	41
Derelict	28	18.5	9	18.5	37
TOTAL COLUMN	175		175		350

Calculate expected values as follows:

$$\text{Expected} = \frac{\text{Total in row} \times \text{total in column}}{\text{Total altogether}}$$

$$*\frac{175 \times 37}{350} = 18.5$$

$$\chi^2 = \Sigma \frac{(O - E)^2}{E}$$

$$= 0.12 + 4.0 + 0.64 + 0.03 + 4.12 + 3.61 + 0.71 + 0.30 + 4.88 + 0.12 + 0.01 + 0.64 + 0.03 + 4.13 + 3.61 + 0.71 + 0.99 + 4.88$$

$$\chi^2 = 33.53$$

Significant at the 0.1% level.
Result of analysis: there is a significant difference between the land use in 1976 and that in 2000.
Next step: explain why.

---

4. On a map of the town draw concentric circles, at 500m intervals, from the central point (PLVI). In each circle use your results from each located point to calculate the percentage of each land use in each ring, for each year of your study.

5. Calculate the percentage of buildings in your study area which were built within the last 5 years / 5–10 years / 10–20 years / over 20 years.

Interpretation and Conclusions

- What was the pattern of land use in each of the years under study?
- How has that pattern changed over time? What evidence is there from the *type* of land use changes to explain why those changes have occurred?
- Has there been more change in some parts of the town than others? What is the nature of that change? Why has it occurred there?
- Were the changes in land use associated with local planning stategies of the time and/or national trends in city and town developments?
- Who has been affected by the land use changes over the years? Have the developments or changes you have identifed been beneficial?
- How may the town change in future? What are the current trends as outlined in the structure plan for the area?

Resources

'Ecology of retailing', Geography Review, Volume 11, Number 3, January 1998
Kelly's Street Directories (until 1974); after that time use Kelly's Business Directory for the appropriate year
Structure plans from planning department – current and former plans
Civic Trust / local history sections of libraries have post-war documents
Goad maps and street plans
Local Chambers of Commerce hold some records of commercial activity in a town

Comparative analysis of a CBD and an out-of-town retail centre

Starting Points

Many urban areas have out-of-town retail centres, and many businesses claim that trade has been affected by these developments. Can you identify and suggest reasons for the differences between the high street and the outlying commercial centre? Has your local retail pattern changed in the same way as others described in textbooks?

It is possible to assess the impact of out-of-town centres by investigating some or all of the hypotheses below. Consider the retailing pattern of a local area and investigate any issues that are particular to your centre.

1. There are differences in the land use pattern between CBD and the outlying retail or commercial centre.
2. The CBD and out-of-town retail centre show distinct differences in type and volume of traffic flow, pedestrian density and shop sizes.
3. Clustering of certain shops occurs in both CBD and outer retail centre.
4. The spheres of influence of CBD and outer retail centre vary significantly.
5. Out-of-own retail centres appear to be taking trade from the traditional high-street shops.

Geographical links to your syllabus

- Concept of bid-rent.
- Town centres are under pressure from competition for space and consequently high land values.
- Nearest neighbour analysis.
- Reilly's law of retail gravitation.
- Spheres of influence are related to the status of a shopping centre.
- Some higher order shops tend to cluster while others do not.

Primary Data Collection

1. Draw land use maps of both shopping centres. Select 30 shops – a systematic sample will ensure that you cover the whole centre equally. Pace shop frontages (to calculate rateable value per metre frontage).
2. Conduct traffic counts (amount and type) and pedestrian flows. Consider and be able to justify the number of counts, the time of day, which day(s) and locations.
3. Select 50 car tax discs in car parks to give an indication of place of origin for sphere of influence. Use random number tables or a systematic sampling method.

4. At the out-of-town centre people don't walk far from their cars so think about how you can record this. You could record multiple visits where people visit shops near or next to each other, particularly if they are comparison shopping.
5. Conduct a shoppers' questionnaire (see figure 39, page 54). You **must** pilot your questionnaire and assess whether or not it will give you the data you need. Beware – you are looking for *explanations* as well as *patterns*. This is more than you considered for GCSE.

Secondary Data Collection

1. Collect rateable values from the local council or Inland Revenue to give an indication of land value per metre frontage, for both the out-of-town centre and the town centre.
2. Find out which shops in the town centre have changed since the out-of-town centre was opened. What has replaced each one? How many shops are now vacant? How long have they been thus? Are there figures for before as well as after the new development?
3. Collect details of the delivery areas from shops in both areas.

Ideas for recording and analysing data ▸▸

1. Draw land use maps of each centre. Annotate to emphasise similarities and differences.
2. Draw a histogram to show shop closures or changes – before and after the out-of-town centre was established.
3. Annotate the map of each centre to show sphere of influence. Add appropriate comment or detail about transport network, main car parking, park-and-ride facilities, which support conclusions about the shape and size of sphere of influence.
4. Use nearest neighbour analysis to identify shop clustering. If you cannot calculate area accurately, find the average distance betwen nearest neighbours for each category of shop. This can then be used as a comparison between shop types.
5. Chi Squared (see figure 38) can be used to compare differences in types of shops in each centre if you group your data (see figure 40).
6. Draw flow line maps for traffic and pedestrian data. If you undertook a stratified sample of pedestrians, are there differences in shopping habits with age of shopper?
7. Use bar and pie charts to graph questionnaire responses. Make sure your graphs are arranged several on one page so that you can compare the results. Graphs need to be annotated, ie summarise in a label what each one shows, with a note about any unusual data.

⚠ *Things to look out for*

Pedestrian and traffic counts are both very tricky so plan very carefully. Town centres are frequently pedestrianised. Counting pedestrians in a traffic-free zone will be difficult.

It is critical that you consider how to select your respondents for a questionnaire. You will not be selecting *randomly*, ie using random number tables. More likely you will be making a *systematic* sample by choosing, for instance, every 10th person. You may decide to identify differences in shopping habits between various age groups so a *stratified* sample is appropriate. For advice about questionnaires (see figure 1, page 3).

Size of sample is also important – it is unlikely that it will be large enough to be conclusive, but collect at least 50 responses from each centre.

Think about which day. Weekdays and weekends are very different as are bank holidays.

Interpretation and Conclusions

- Summarise the types of shop in each centre.
- Identify any differences in shopping habits, land values and pedestrian flows. Use information from the questionnaire to provide or suggest explanations. Is the age of the shopper significant?
- Do shops in both centres cluster? Why? Is the reasoning the same for both shopping centres?
- Are the spheres of influence different? How? What factors influence the shape of each sphere of influence?
- How and why do traffic flows differ between each centre? Are there differences between weekdays and weekends? Why?
- Bearing in mind your conclusions, can you evaluate the impact of out-of-town retail centres on traditional town centres?
- What further investigations could be undertaken to follow up your enquiry so far?

Figure 39 Potential questions to shoppers

Where do they live?
How did they travel?
How often do they shop here, and what type of goods purchased?
Do they visit other shops at the same time?
Do they visit the other shopping centre in the study, if so how often, what do they buy, and how do they travel?
What other shopping centres do they visit?
Can they comment on attractions of shopping in a particular centre – convenience, transport routes, parking, choice, atmosphere, ease of travel etc.

Figure 40 Table of shopping centre differences

TYPE OF SHOP	NUMBER IN TOWN CENTRE	NUMBER IN OUT-OF-TOWN DEVELOPMENT
Department store	9	0
Clothes	41	1
Electrical	2	5
DIY	1	3
Specialist comparison goods	56	1
Leisure / Food	23	5
Other	19	1

Resources

Chapter 6 of *Human Systems and the Environment* by R Prosser, Nelson, 1992
Developers who establish out-of-town retail centres have maps and materials and some have websites

Comparative analysis of the socio-economic characteristics of different areas in a town

Starting Points ▶

1. Many areas in towns and cities display distinct social and economic characteristics.

2. These differences are also reflected in house type and value.

3. There are significant differences in environmental quality across a town.

4. Some areas experience more change in population and land use than others.

Geographical links to your syllabus

- Land values vary according to accessibility. Bid-rent theory.

- House type and value and environmental quality are a product of the historical evolution of towns.

- Changing patterns of work are reflected in changing patterns of mobility across a town or city.

⚠ *Things to look out for*

Selecting your areas can be difficult and subjective. Wards are large areas and usually consist of a variety of house types to ensure fair representation at local government level. You will find it difficult to make a 'typical' EQA from one street. You may decide to work on ED areas (Enumeration District areas) which consist of about 200 households. This will give you a much more focused study. EDs are listed by number in the Census, but are not normally identified by area. You will need to find a helpful planning officer to locate precisely where each ED is. The information could be regarded as sensitive, so you need to be very careful how you approach the analysis at this level of detail.

You need 10 areas in order to test any relationships using Spearman Rank or Pearson's Product Moment coefficient.

Primary Data Collection

1. Select 10 areas in your town (see notes below). For each one choose a street which you consider is typical of the area. Be prepared to explain your choice of location in your writing up.

2. Conduct an environment quality survey of each area (see figures 33 and 44).

3. Identify the house value index for each ward by calculating the average house price. Use the local papers to collect house prices. Take the average value from five detached, five semi-detached and five terraced houses.

4. Conduct a household security survey. Observe whether a selection of 20 houses in your ward has a burglar alarm, security lights, gate, intercom and surrounding wall.

5. Take photographs or draw sketches of the area to identify their characteristics, including building age, general appearance, litter, graffitti, parked cars, evidence of care or decay.

6. Use a bi-polar exercise to quantify your views of each area. Ask someone else to complete the exercise too (see figure 41). It would be very interesting to ask someone from each area to assess their own locataion, and discuss the effect of residence on people's views.

Secondary Data Collection

Use the 1991 Census to collect data on your selected wards. Some local councils have collated a local atlas from their Census data which you may find easier to use. Note figures for: socio-economic groups; levels of employment (percentage of unemployment in area); ethnic origin; population mobility (number of people who have moved into the area within the last five years); car ownership; percentage of housing with no central heating (a measure of housing quality and wealth); housing density; percentage of housing with more than one person with long-term illness (a measure of health and social pressure); population structure of each area. Choose indicators which you feel are particularly relevant to your area.

Does a shopping or service hierarchy exist in an urban area?

Starting Points ▶

1. Are there shopping centres of different sizes in a city?
2. How do the catchment areas of the shopping centres of different sizes vary?
3. Do shopping centres provide different functions?
4. Do shopping habits vary between centres of different sizes?

Geographical links to your syllabus

- Just as locations in cities vary with age, house type and density, and income, so local shopping centres vary accordingly. There is a hierarchy of shopping provision in cities.

- Small shopping centres provide low-order goods and services for small areas whereas larger centres provide higher order goods with a larger catchment area. Indicator functions can be used to identify different levels in the shopping hierarchy.

- Changes in shopping patterns have had significant effects on the patterns of shopping in cities.

⚠ *Things to look out for*

If you include corner shops as first order shops you may have problems collecting data. You should be able to collect more data from the town centre – the highest level of shopping in an area.

Select your sampling method and timing for your questionnaire with care. Local shopping centres support different populations at different times of the day and week.

Justify your sample size.

Figure 46 Indicator functions

These could include some or all of the following: general store, newsagent, butcher, Spa / supermarket, bank, building society, chemist, dry cleaner, travel agent, estate agent, large supermarket, ladies' clothes, sports shop, stationery shop

Primary Data Collection

1. Select at least 15 different shopping centres of different sizes from around the large town or city.
2. Select 10 indicator functions which represent different levels in the urban hierarchy (see figure 46).
3. Count the number of each of the indicator functions at each centre and the total number of shops.
4. For each centre map the shopping provision.
5. Conduct a questionnaire at each centre to find the following:
   - What do you buy here?
   - How often do you shop here?
   - How do you travel to get here?
   - Do you shop elsewhere in the city?
   - If so, what do you buy there?
   - How often do you shop elsewhere?

Secondary Data Collection

1. Collect rateable values for the indicator functions in each of the shopping centres.
2. Use the Census to find population size of each settlement.

Figure 47 Calculation of the centrality index

Centrality value is the inverse of the number of times a function occurs across your whole fieldwork area.
- Divide the **total** number of, for example, chemists, you have counted in *all* your fieldwork. Divide this number (n) into 100. **This is the centrality value for chemists, eg CV = 100/n**
- Repeat this for every indicator function you are working on.
- For every settlement, work out the total score of centrality values. If there are 2 chemists in the settlement, multiply the CV number by 2. If there are 6 banks, multiply the CV number by 6.
- Add this total score of centrality values to the total number of shops in the settlement. The final figure is the **centrality index.**
- Rank the shopping centres using the centrality index.

Figure 41 Bi-polar semantic differential scale

+2   +1   0   −1   −2

High-density houses	Low-density houses
Congested	Little traffic
Area poorly looked after	Area well looked after
Low property values	High property values
Lack of open space / gardens	Plenty of open space / spacious gardens
Mainly on-street parking	Cars off street when parked
Noisy	Quiet

Give each factor a score from +2 to −2. Think about the factors you wish to identify. These are some examples only. There may be others appropriate to your town. Add up the total score for each area and compare your scores with those of other people and for different areas.

Ideas for recording and analysing data ▶▶

1. Draw choropleth maps of all your data across the wards. This will show differences from the centre to the edge of the urban area.

2. Graph several variables on a line graph to identify any relationships between them. Put the different areas along the x axis. Annotate the graph and cross-reference to the sketches or photographs which will illustrate the points you are making.

3. Overlay the house value choropleth map with a map of the route network to identify links between land values and accessibility.

4. Draw population pyramids of each ward. Annotate to show differences or similarities. Put several on one page.

5. If you wish to investigate relationships between variables, use Spearman Rank or Pearson's Product Moment to correlate. Beware – just because you find a correlation does not mean cause and effect.

6. Calculate Z numbers (see figure 42) to quantify the standard of living of an ED or ward. This will enable comparisons to be made between areas.

Figure 42 Using Z scores

Decide on the scale of area to be investigated, ie Enumeration District or ward.

Take a selection of indices which give an indication of deprivation such as:
- percentage of housing with no central heating
- percentage of housing with at least one person with a long-term illness
- percentage unemployed
- percentage of houses with no employed adult
- percentage of houses with more than 1.5 persons per room
- percentage of people in socio-economic group V

For each index, list the data for every area as a percentage.
Find the mean value for all the areas.
Calculate the standard deviation for this data.
Subtract the mean from every area value.
Divide the result by the standard deviation.

This gives you a Z number for that particular index for every area. You can now compare numbers for different areas. If the number is positive it means that the indicator in an area is above the average. If it is negative the indicator is below average.

Calculate Z numbers for all the indices you have selected. Add the Z numbers to give a total Z score for all variables for every area.

---

7. Calculate and map the location quotient, eg

   $$\frac{\% \text{ rented accommodation in an ED / ward}}{\% \text{ town's total housing in the ED / ward}}$$

   This figure is calulated for all areas, then mapped on a choropleth map.

8. Draw a bi-polar diagram for each area or contrasting areas. Compare your views with those of other people.

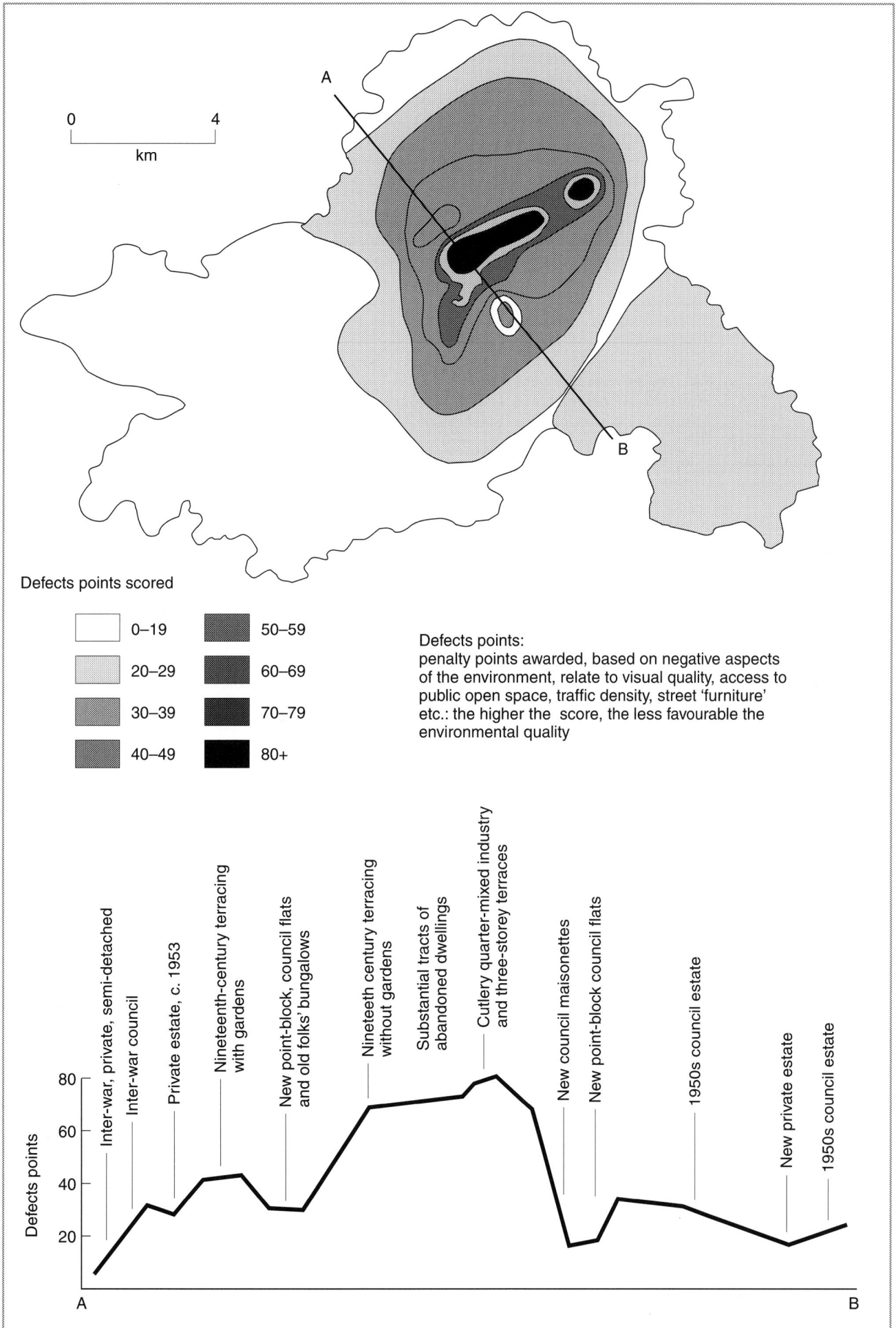

Figure 43 Map of environmental quality in Sheffield

Figure 44 Environmental quality survey

		Points
Land Use	Exclusively residential	O
	Some non-residential uses	1–3
	Majority non-residential	4–5
Parking	Provision for all cars to be parked off street	O
	Some off-street parking	1–3
	No off-street parking	4–5
Landscape quality	Mature trees; well-kept, plentiful grassed spaces	O
	Few trees; poor-quality, unkempt grassed spaces	1–3
	Total, or almost total, lack of trees or grassed spaces	4–5
Built environment	Attractive	O
	Some drabness	1–3
	Very drab	4–5
Garden provision	All houses with adequate gardens	O
	Communal open space or tiny garden space	1–3
	No gardens or communal open space	4–5
Garden appearance	Tidy, well screened, cared for	O
	Some poorly cared for, poor screening or fences	1–3
	All gardens poorly screened or fenced; overgrown	4–5
Traffic	Residential traffic only	O
	Some intrusion of through traffic or unsuitable trafffic	1–3
	Substantial intrusion of through traffic	4–5
Noise	Acceptable residential standards; normal speech possible	O
	Slightly above acceptable residential standards; some interference with normal speech	1–3
	Above acceptable residential standards; normal speech difficult at times	4–5
Air pollution	Negligible	O
	Light – some staining of walls and buildings	1–3
	Heavy – severe staining of walls and buildings	4–5

Interpretation and Conclusions

- Are there differences between wards? For all variables? Some variables? Do some wards show the same trends over a number of variables? Why?
- Are these differences related to distance from the town centre? Are they related to building age?
- How does environmental quality change across the town (see figure 43)? Is it what you expected? Why? Are there any exceptions? Why?
- Are patterns of employment related to mobility of the population, accessibility of each area, house values, and age structure of the population?
- Do physical factors of site or situation explain the patterns you have identified?
- How do theoretical ideas relating to the different variables apply to your urban area?
- Is your area typical of other British towns?

Resources

'Deprivation and the 1991 Census', Robin Holmes, *Geography Review*, Volume 8, Number 3, January 1995
1991 Census on CD-ROM; most libraries will have a copy

Analysis of the sphere of influence of a town or village

Starting Points ▶

1. Do different facilities in a town have different spheres of influence? Why?
2. What factors influence the shape of the sphere of influence of individual services?
3. What is the widest extent of influence of the town?
4. What factors influence this wider sphere?

Geographical links to your syllabus

- Christaller's ideas on size of market area and type of services provided.

- Similar order of services should supply similiar sized service areas.

- Other geographical factors influence the shape of spheres of influence, notably physical features, transport facilities, personal preference, and quality of alternative facilities.

- Distance decay may reduce the strength of influence with distance from the town centre.

⚠ Things to look out for

Sample size. See figure 1 concerning questionnaires.

Sampling method: you may consider a stratified sample for your leisure sphere of influence, i.e. those people in particular age groups. A systematic sampling method, i.e. every 10th person, might be appropriate for your shoppers' questionnaire. Similarly a systematic sample could be used to collect tax disc data from parked cars.

Primary Data Collection

1. Identify a number of contrasting services provided by your town or village. Include different functions, eg administrative, leisure, transport, education, and different orders of shopping, such as newsagents, furniture, tourist, local newspaper, estate agents.
2. For each type of function find the furthest limits of their activity, ie their sphere of influence. Some will be easier than others. You should visit each function to map their boundaries.
   a) The local authority administrative boundary is available from the council offices. Plus Police, Fire and Ambulance services boundaries.
   b) Transport – the limit of local bus services.
   c) Identify the catchment areas of the local secondary schools.
   d) Leisure – conduct a questionnaire of leisure centre users to find out where they live (first part of their postcode) and how they travelled to the leisure centre. Think about how many people to sample, how to choose respondents, and how often and when to visit, in order to cover the range of activities on offer at different times during the week.
   e) Car parks – car tax discs indicate place of taxation and you can make an assumption that is near the place of residence.
   f) Visit all the newsagents in the town armed with a street map and plot the limit of their newspaper deliveries. Also try to find out what proportion of their sales comes from passing trade.
   g) Visit some major stores who deliver goods. Find out the limit of their delivery areas, including the sliding scale of payments for deliveries over further distances.
   h) Conduct a shoppers' questionnaire to find out where people who shop in your centre live, and how they travelled there. Think about sample size and sampling method. It may be appropriate to look at the spheres of influence for young people under 21, those in the 22–60 group, and those over 60.
   i) Can you discreetly find out where people work?

Secondary Data Collection

1. Using an Ordnance Survey map, identify potential physical factors which could impact on spheres of influence – topography, rivers, bridging points, tunnels, areas of forest with low-density population.
2. Using the local newspaper, map the areas covered for local news items and the location of commercial and private advertisers – use the telephone codes from the 'Sales' sections of the paper. Can you work out what volume of advertisers come from different places? This will be more effective for a town than a village.
3. The Census has details of travel to work areas.
4. Collect details of areas of telephone codes from the Postal Address Book available in local libraries.

Ideas for recording and analysing data　▶▶

1. Draw a base map of the key physical features within the wide area of influence you have identified. Don't forget a scale.
2. Map the principal transport routes on a base map. Include key bus, road and rail routes.
3. Using these base maps, draw the spheres of influence for each type of service or function you have investigated – a separate map for each one. Annotate them to draw attention to particular points about their shape or unusual features.
4. Superimpose the spheres of influence on to one map using different colours.
5. Draw scatter graphs of number of shoppers *v* distance travelled. If there appears to be some relationship, use Spearman Rank to correlate. Don't forget to test for signficance – has this relationship occurred by chance?
6. If you conducted a large enough questionnaire survey you can investigate similar correlations using different age groups.
7. Use Chi Squared to test whether or not there is a significant difference between the number of newspaper advertisers or news articles in areas near the town compared with areas further away.

Interpretation and Conclusions

- What are the differences in spheres of influence as shown on the superimposed map?
- Can differences be explained in terms of the nature of the functions or services?
- Is there a distance decay effect? Why or why not?
- Does age have an effect on the shopping sphere of influence?
- Does type of leisure activity affect the sphere of influence for leisure?
- Do physical features appear to have an influence on the shape of the spheres of influence? How? Why or Why not?
- Do commercial spheres of influence match administrative, leisure or retail spheres? Why or why not?
- In the wider context, would it appear that the spheres of influence you have mapped overlap with any other areas? Why? How could you follow this up?

Resources

1991 Census
Town website
Postal Address Book produced by the Royal Mail will give you maps or details of postcode locations

Comparison of retail spheres of influence of neighbouring towns

You could compare shopping areas within a town or city, or compare one town with another.

Starting Points ▶

1. Are there differences in shopping facilities between the selected places?

2. How do facilities, services or functions of the towns affect their spheres of influence?

3. What determines the shape of the sphere of influence?

4. How do other factors such as topography or transport network affect the sphere of influence?

5. To what extent do spheres of neighbouring towns overlap? Why?

Geographical links to your syllabus

- Reilly's Law of Retail Gravitation can be used to identify the theoretical trade limits of towns.

- Settlements can be ordered according to the functions or services they provide. Higher order settlements generate larger spheres of influence.

- Places of similar size and with similar shopping provision could be expected to have similar sized spheres of influence.

- The precise shape and size of the sphere of influence depends on other geographical factors, such as transport network, administrative boundaries and personal preference.

⚠ Things to look out for

Choose towns of equal sizes so that comparison can be more objective.

Justify your comparison. Adjacent towns? Competing towns? Towns with similar functions?

If you are looking at shopping centres within a town or city, make sure they are independent from one another.

Pilot your questionnaire carefully.

Sample your shoppers in the same way in each place. Remember that populations differ in different towns. This may affect the sphere of influence. You may need to refer to the population structure of each place.

Primary Data Collection

1. Conduct a shopping survey in each town centre. Remember that this has to be thorough. Investigate the side streets as well as the main shopping area. Count the number of outlets in each category, eg finance, food, department stores, clothes, specialist shops, electrical etc.

2. You could be very sophisticated and create a weighting for shops of different sizes, eg to differentiate between small food shops and major supermarkets, or between factory outlet clothes shops and major department stores.

3. Evaluate the shopping area by classifying its characteristics of shop quality and street appearance. This is subjective but may help you compare places (see figure 45).

4. Conduct a questionnaire of 50 shoppers in each town. Think about the following questions. Make them closed questions, ie give a series of options for the answers. It will prevent rambling responses.
Where do you live? How did you get here? What do you usually buy when you come here? Do you ever visit town A or B? If so, how often? What do you buy? What influences your choice of shopping centre?

5. In the car parks use car tax discs to record the home area of visitors. (Of course, cars do not necessarily belong to shoppers.)

6. Record other administrative functions of each town.

Secondary Data Collection

1. Use OS maps to identify topography, rivers or other features which could influence the shape of the sphere of influence of each town.

2. Map the administrative boundaries of each town. Identify limits of transport services, primarily key bus routes, main roads and railway links. Note where bus routes run from one town to another.

3. Try to establish from bus companies how many passengers travel from one town to another, and if numbers vary during the week.

4. Identify the population structure for each town.

Figure 45 Classification of shopping quality and street appearance

Score on scale of 1 to 5

Shopping quality

	PLACE 1	PLACE 2	PLACE 3	PLACE 4	PLACE 5
A Type of shop 1 = area dominated by department stores or shops selling comparison goods 5 = a variety of shops – mainly convenience shops					
B Other land use groups 1 = mainly shops 2 = banks and shops 3/4 = mainly offices 5 = mainly other uses					
C Retail organisations 1 = national chain stores dominant 3 = $\frac{1}{2}$ national $\frac{1}{2}$ independent 5 = independent dominant					
D Quality of goods 1 = good quality / high price 5 = poor quality / low price					

Street appearance

E Street crossing safety 1 = very safe 3 = busy with crossings 5 = very busy with no crossings					
F Shopping crowds 1 = dense 5 = sparse					
G Cleanliness 1 = very clean 5 = dirty					
H Shop appearance 1 = very nice; good window display 5 = poorly maintained					
I Traffic to pedestrian 1 = pedestrianised area 3 = open to traffic 5 = main traffic route – no parking					
J Vacant properties 1 = all occupied 5 = majority vacant					

Ideas for recording and analysing data ▶▶

1. Draw land use maps for each town, using the same scale and land use categories.
2. Tabulate the number of each type of shop provided in each town. Draw a divided bar chart to show proportions of each shop type in each town.
3. Use Chi Squared to test whether or not there is a significant difference between shopping provision in each centre. (You may want to show there is no difference in shopping provision in each place.)
4. Draw flow line map(s) to show volume of shoppers from different locations.
5. Draw scatter graphs of distance v number of shoppers for each town. Test correlation using Spearman Rank.
6. Map the transport network around each town to identify the types of road. If rail transport is important, you need to consider that too.
7. Draw flow line maps to show bus routes and volume of services. Identify particularly the end of the bus routes.
8. Graph the responses to your questionnaires. Remember that you are comparing, not investigating one town at a time. Put several graphs on one page. Use a colour code for each town. Annotate the graphs to show the main features or the differences that you will be writing about.
9. On a topographic map annotate the physical features which affect the transport routes.
10. Map the administrative boundaries for each place.
11. Map the place of residence of shoppers to each town.
12. You can use overlays to compare each place but beware of putting too much information on one map. If you keep each map free of clutter it will be easier to analyse.

Interpretation and Conclusions

You are doing two things here. First, you should draw together data for each town to identify and explain the size and shape of the sphere of influence of each place. For instance:

- How far do people travel to shop in the town? Does the transport network have an effect? What about topography?
- Briefly summarise the shopping attractions of the town. Does it cater for a particular group in the population? To what extent do shoppers visit the other centres, and why?
- Are there any particular functions in the town which are important at a regional scale rather than a local scale? How does this affect your results?

Second, you need to compare each town; as follows:

- How do the spheres of influence vary? Why? Consider the importance of transport or topography in each case.
- Is there a significant difference between shopping provision in each place? Can you explain why?
- Does the administrative boundary have any effect? Why or why not?
- To what extent do shoppers visit the selected places? Explain your results.

Finally, are your results what you expected to find? Can you make some generalisations about spheres of influence and the types of towns you have been investigating? Would you expect your results to be replicated elsewhere?

Resources

'Ecology of retailing', *Geography Review*, Volume 11, Number 3, January 1998
Town websites
Goad maps of towns
Shop advertisements in local newspapers will also give a version of sphere of influence

Investigation of the relationships between sphere of influence and settlement size

Starting Points ▶

1. Larger settlements usually contain higher order functions.
2. People travel further to use higher order functions so the sphere of influence of larger setlements is proportionately greater.

Geographical links to your syllabus

- Central Place Theory, the Christaller model, which describes a hierarchy of settlement size and market area.
- Concepts of range, market area and threshold population.
- Sphere of influence depends on topography and transport networks.

⚠ *Things to look out for*

Your selection of settlements is important. You need a range of population sizes, and you need to be able to visit each one easily. They should be clearly identifiable and separate from each other but within a clearly defined area, eg 75–100km².

Primary Data Collection

Select a number of settlements of different sizes (at least 10 places). Make sure they are 'discrete' places, rather than suburbs of cities or conurbations.
1. For each settlement identify the highest order retail function.
2. Conduct a questionnaire of 50 people in each place using the highest order function to find out where they live and how they travelled.
3. Identify the delivery area and advertising area of the highest order retail function in each place in your study area.

4. Apart from retailing, list other major functions of any of the settlements you are studying.

Secondary Data Collection

Using the Census, identify the population size of each place you are investigating.

Interpretation and Conclusions

- Do spheres of influence vary in size? What influences their shape?
- Is there a correlation between settlement size and sphere of influence?
- How do your results compare with the theoretical patterns of geographical models?
- How might functions other than retailing impact in the sphere of influence of a settlement?
- Is there anything about the distribution of settlement in your region which could have an impact on its sphere of influence, for instance, tourist functions or a glaciated upland with dramatic topography?

Resources

Town websites
Web sites of other non-retail functions or organisations located in any of the settlements

Ideas for recording and analysing data ▶▶

1. For each settlement use the place of residence data to draw a map of the sphere of influence of the highest order retail function.
2. Calculate the area of each mapped sphere of influence.
3. Identify topographical and transport features which could affect the shape of the sphere of influence.
4. Graph the travel data collected from each set of questionnaires to link the most important mode of travel to the map of topography or transport network.
5. Identify a hierarchy of settlement size.
6. Draw a scatter graph of settlement size *v* size of sphere of influence. If the relationship appears to be a linear one test the correlation using Spearman Rank. Test your results for significance.

Ideas for recording and analysing data   ▶▶

1. Map the distribution of settlements and represent population size. Proportional circles are a typical method.
2. Draw small land use maps of each shopping centre and annotate main characteristics such as parking facilities and public transport access. Remember to include a scale in order to compare the size of shops between centres.
3. Use a variety of graphical techniques to represent questionnaire responses. Remember that these need not be large – several on one page, coordinated with retail land use map will enable you to link data together.
4. Calculate the centrality index for each settlement (see figure 47).
5. Using semilog paper, graph centrality index on log scale against rank of shopping centres on the linear scale. Can you identify different groups of settlements on the graph (see figure 48)?
6. Draw a scatter graph of rank of population size *v* rank of centrality index. If there appears to be a linear relationship, test using Spearman Rank, and a significance test.

Interpretation and Conclusions

Describe the pattern of different types of retail setttlements in your area. Is there a hierarchy? Why or why not? Remember – whatever pattern you identify needs an explanation. Consider a variety of specific local factors which could influence settlement hierarchy such as socio-economic status; dormitory function; proximity to alternative centres; quality of transport network / by-pass / rail station / motorway junction; existence of a regional or national function or tourist attraction.

What is the relationship between hierarchy and transport network? Are larger retail centres located on major routeways and neighbourhood centres within housing estates? Try to compare the distribution you find with theoretical ideas.

Resources

Introducing Towns and Cities by K Briggs, Hodder & Stoughton, 1974

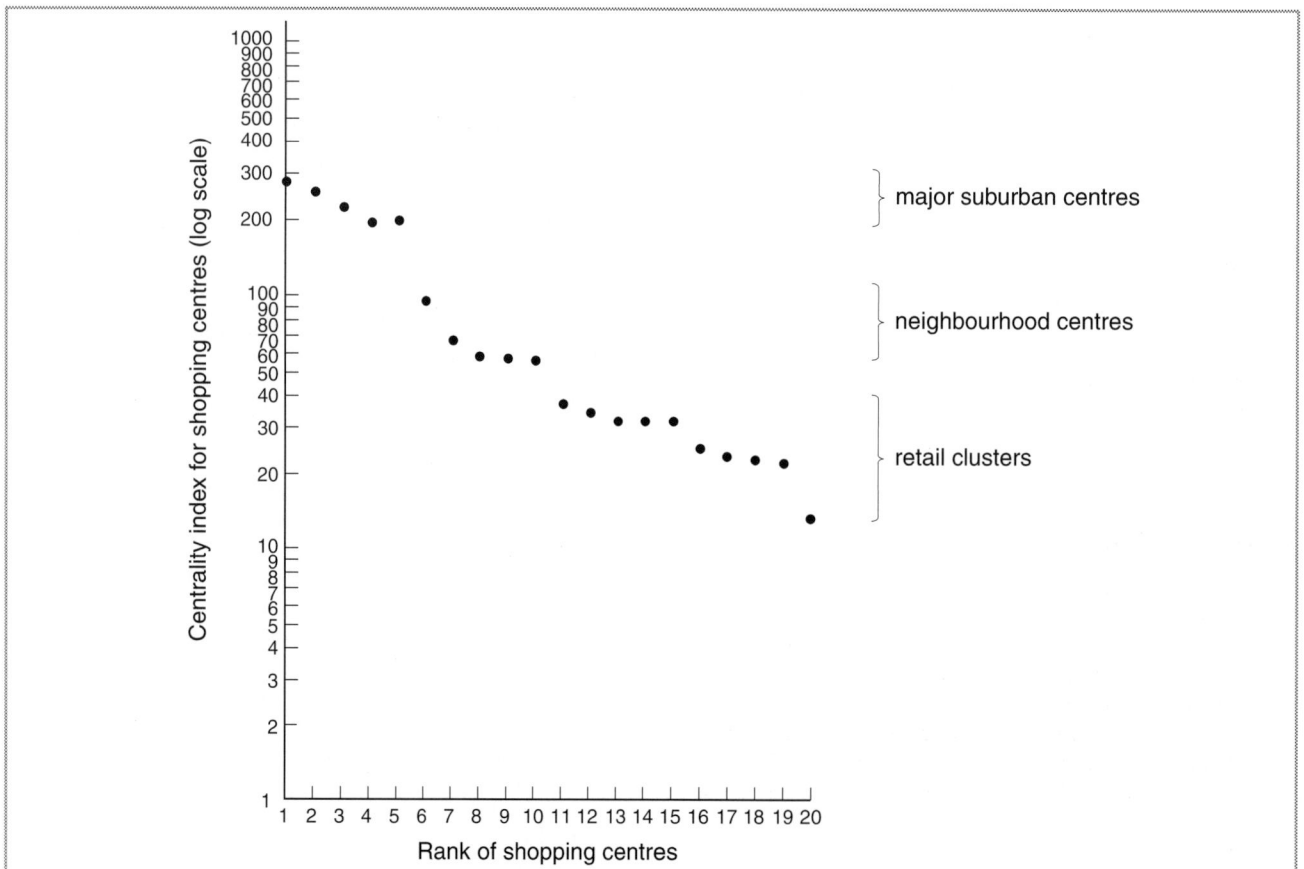

Figure 48 Scatter graph of shopping centres in an urban area

Analysis of a suburbanised village

Many villages are now commuter settlements and their character has changed considerably. Your fieldwork would be to assess the extent to which that is true of your village.

Starting Points ▶

1. There have been substantial developments and changes in housing styles and housing density in the village.

2. There have been significant changes on the character and structure of village population; residents' work and patterns of employment; and the character of services provided in the village.

3. There is increased noise, pollution and lack of open space.

4. A close-knit community has been replaced by a larger, more divided community.

Geographical links to your syllabus

- The process of counter urbanisation has a major impact on rural settlement.

- Suburbanised villages have a distinct character in the early twenty-first century.

Primary Data Collection

1. If the village is small enough, make a detailed land use map of all the housing, and note age, type of house, building materials, housing density and street plan layout. If it is too large you will have to sample streets from each stage of development in the village. Your pilot study will identify which areas need to be sampled.

2. Conduct traffic counts on the main roads leading into the village at 8–8.30am and 5.30–6pm, ie commuter traffic. Note amount and type of vehicle. Compare with traffic counts and type at different times of the day.

3. Using a stratified sample, conduct a questionnaire of at least 100 residents (see figure 49). If you know the proportion of different house types, this can be the basis for your sample since house type is a reasonable indicator of wealth.

Secondary Data collection

1. Detailed data at ward and ED level is available from the 1891 Census and the 1991 Census. Intervening Censuses provide demographic data at urban or rural district level. Occupational data is grouped at county, municipal or county borough level only. Old parish registers give details of births, marriages and deaths with occupations, but sampling this material could be long and onerous.

2. Photographs from historical books about the village will identify village shops, schools and leisure facilities as well as local businesses.

3. Obtain street maps of the village at various times in the last 200 years.

4. Obtain a current street plan from local estate agents.

5. Kelly's Directory lists every business in each village. This is available until the 1980s. After that time use Yellow Pages to identify what businesses there were in the village. Some counties also have Thomson Directories.

6. Find out the frequency and destination of bus and train connections.

⚠ *Things to look out for*

Make sure this does not become a historical record of the village.

Be aware that villages can be closely knit communities where people may not be prepared to give you all the details you would like because of confidentiality.

If you are making extensive use of Census data make sure you are familiar with boundary changes which have periodically taken place.

Figure 49 Example of a questionnaire

It should identify the following information:
House type: detached / semi-detached / terraced / cottage
Building type: brick less than 200 years old / stone / cottage / thatched over 200 years old
Length of residence in village: less than 5 years / 5–10 years 11–20 years / over 20 years
Age of main occupant: 18–34, 35–59, 60+
Children in the house: yes / no
Main occupant is / was / will be involved in: manual / professional work
Workplace: commuting over 10kms (give location) / work locally
Type of industry where residents work: extractive (forestry, fishing, mining) / manufacturing / service / high-tech industry
Car ownership in the household: no car, use public transport / regular use of 1 car / regular use of more than 1 car
Description of local community: small, close-knit / divided socially into smaller community groups / isolated
Noise and environmental pollution: a lot / a little

Ideas for recording and analysing data ▶▶

1. Graph the population growth of the village. Draw and annotate the population pyramids for the population in 1891 and 1991. Calculate the dependency ratio of the population at different times during your study period. How has the character of the population changed with social trends?

2. Make a detailed map showing changes in land use over the years. Annotate to link with the photographs and observations you have made about building materials and house styles.

3. Graph the questionnaire data and annotate the features which indicate evidence of a commuter function for the village. Identify what proportion of residents are new to the village. Link the growth with evidence of building housing estates.

4. Draw a flow line map of commuting patterns in and out of the village in the morning and evening. Use Chi Squared to test whether or not there is a significant difference between traffic at each time.

5. Map and date the businesses, factories and offices in the village. Identify the number of people employed in each one.

Figure 50 Is there a significant difference between morning and evening volume of traffic?

	TRAFFIC ENTERING	TRAFFIC LEAVING	ROW TOTAL
8–8.30am	18	39	57
6–6.30pm	27	15	42
Column total	45	54	99

$\chi^2 = 10.41$. Significant at 1% level.

Result: there is a significant difference between the number of cars entering and leaving the village in the morning and evening. This confirms the hypothesis and suggests that this is a commuter village.

6. Map the road network, bus and train routes. Annotate to identify the most important routes and suggest reasons for the pattern.

Interpretation and Conclusions

- Summarise the physical growth of the village over the years.
- How have employment patterns changed? How has the age structure changed?
- Identify and analyse the evidence that the village is a commuter-based settlement. Suggest why the village has evolved in this way.
- To what extent do you think the village is now more of a divided, larger community?
- To what extent does your village conform to the theoretical patterns of a suburbanised village (eg models by Waugh and Hudson).

Resources

Local Civic Trust or local history group
Local structure plan for the nearest town

Recreation pressure in a Country Park or Forest Park

Starting Points ▶

1. What types of recreation take place in the Country / Forest Park?

2. Some leisure activities result in more pressure and misuse than others. Are the recreation activities compatible or is there conflict? If so, what is the evidence?

3. There are differences in visitor numbers to a Park on a daily, weekly and seasonal basis.

4. The number of visitors declines with distance from the Park.

5. Differences in visitor numbers are related to types of leisure activities undertaken.

6. The impact of tourists decreases with distance from car parks.

7. What management strategies are adopted to manage different demands made on the Park?

8. Management of leisure resources is most effective where visitors have information and understanding of the issues involved.

Geographical links to your syllabus

- Country Parks were set up to provide areas solely for leisure activities, and located within easy reach of urban areas.

- Parks therefore face pressures of over-use and conflicting demands from different users as a result of being close to areas of dense population.

- The effect of distance decay influences the origin of visitors to a Park.

- Access and accessibility are important issues in management of Parks.

- There is a close relationship between conservation, education and the quality of outdoor leisure activities.

Primary Data Collection

1. Identify the range and location of leisure activities undertaken in the Park.

2. Identify through observation the various management strategies used in the Park.

3. Interview the Park Warden to find details of management strategies; areas of the Park specifically targeted and why; the length of time that management strategies have been in operation; environmental problems which existed before the management was established; limitations of the strategies used; and the effectiveness of those strategies from the Warden's view.

4. Conduct a questionnaire of visitors to discover frequency and times of use; how long visitors stay; distance travelled; how they travelled; type of leisure activity involved. Also ask about the role of the Visitor Centre and visitors' views on management methods within the Park. Consider a stratified sample if you wish to investigate leisure activities by different age groups. *Remember – be able to justify your sample size and method.*

5. Observe origin of cars in car park using car tax discs of 30 cars in each car park. Remember that this will be an approximate indication only.

6. Record amount of litter in 10 areas with increasing distance from car parks, popular sites and / or litter bins. Count number of pieces of litter in an area 5 × 5 metres.

7. Observe population pressure at various points within the Park. Select 10 footpaths on a systematic basis with increasing distance from car parks. At each site record number of pedestrians, cyclists and riders over a 30-minute period.

8. At each site measure footpath erosion. Stretch a tape across the path. Make sure it is horizontal. Measure distance from tape to ground at 10cm intervals across the path. Alternatively use the footpath erosion survey (see figure 51).

9. Record type, height and frequency of vegetation across each footpath using a quadrat. (Not all paths will have vegetation on them. You should ask yourself why.)

10. Measure compaction across each footpath site using a soil penetrometer.

11. At each site make annotated photographs or sketches to use in your discussion. Use a quadrat in the centre of each path to estimate the characteristics of each path. Identify the percentage of trampled vegetation, bare ground, grass, mud, and bare roots.

12. Conduct an assessment of the effectiveness of the Park management at each site, using bi-polar analysis.

Figure 51 Footpath erosion survey

The following seven main features of erosion are used to categorise a path as it is surveyed.

- **Width of scarring**
  Area over which the vegetation has been lost.
- **Depth of scarring**
  Measurement of soil loss from the original vegetation level.
- **Width of worn vegetation**
  The vegetation adjacent to the path may differ from the surrounding area. This often shows as a precursor to more serious erosion.
- **Braiding present**
  Evidence of the creation of multiple paths.
- **Pigeon holing present**
  Flattened toe holes are a good indicator of vegetation and soil structure breaking down.
- **Gullying**
  Major indicator of serious erosion.
- **Loose debris covering adjacent vegetation**

  At each site record a value from 1 to 5 for the seven main features of erosion.

Source: 'Repairing upland paths erosion: a best practice guide', Erosion recording form, page 2; RFI: 2/2.

Secondary Data Collection

1. Country Park Management Plan will outline the context for any management within the Park.
2. Use any footpath, vegetation or visitor data from earlier investigations. Make sure your primary data is comparable with any gathered from earlier studies.

⚠ *Things to look out for*

Remember – your personal safety *always* comes first.

Make sure that your observations are comparable with any collected by other agencies, ie the Country Park authorities or the local council.

The availability of path data or people-counts could influence your choice of recording sites.

Justify the length of time needed for your people-count at each site. They need to be as simultaneous as possible.

Interview with a warden: you need some specific questions. Prepare yourself well. Be organised. Make clear notes. Identify the main issues of concern, the warden's view of those issues, and which aspects are of greatest concern to him/her/the local authority. Do not take too long. Wardens are busy people, and you will not be the only student seeking information.

Figure 52 Assessment of management strategies

	Very good	Good	Satisfactory	Poor	Very poor
Provision of information along paths					
Provision of information in Visitor Centre					
Physical barriers to path width					
Artificial path surfaces					
Restriction of visitors to certain areas					
Restriction of activities to specific areas					

Figure 53 Environment quality survey

	POOR/SUBSTANTIAL		GOOD/VERY LITTLE	
	1	3	5	
Litter/dirty				Clean
Damage to vegetation				No damage
Noise				Quiet
Crowded				Few people
Unkempt paths				Natural / well-managed paths

Ideas for recording and analysing data ▶▶

1. Draw an annotated map to show location of principal leisure activities in the Park, and the location of your observation sites.
2. Draw a flow line map to show volume and origin of visitors.
3. Use a star diagram to represent volume of people engaging in different activities in the Park; and similar diagrams to show seasonal differences. Chi Squared analysis can be used to investigate differences – if your data set is large enough.
4. Graph the questionnaire data – bar, pie chart etc. Think about how you arrange your graphs. Several on one page allows interrelationships to be recognised.
5. If you have a sufficiently large sample size you can make an in-depth analysis by type of leisure activity or age group, to identify patterns or contrasts between different pursuits.
6. Use flow line maps to represent visitor frequencies throughout the Park.
7. Draw an isoline map of density of litter. Focus on car parks, popular sites and / or litter bins.
8. Draw cross sections showing depth of each footpath, and a kite diagram above to show change in vegetation across the paths (if there is any).
9. Draw a scatter graph of environmental quality survey (see figures 44 and 53) values v distance from the car park(s); also number of visitors v distance from car park(s). If there appears to be some sort of pattern follow up with a correlation test. Remember that any relationships must have some logic and reasoning behind them.
10. Tabulate information from the Park Warden. Sketch the environmental management strategies used within the Park.
11. Construct graphs to represent the effectiveness of management techniques at each site.

Interpretation and Conclusions

- Describe and explain the pattern of leisure use in the Park.
- Is there a distance decay effect of visitors to the Park? Can you map the catchment area? What influences its shape? Refer to OS maps to comment on topography and communications. How important is accessibility in both use and management of the area?
- Are there any differences between different types of leisure activities?
- Is there evidence of environmental pressure in the Park? Is it related to particular types of activities? Does it lead to conlict between different users? Why?
- If you have available data, what changes have occurred since the earlier data was collected?
- How does environmental pressure vary throughout the Park? Is it related to distance from car parks? Is this consistent from all car parks? Why / why not? Do popular locations within the Park create environmental pressure?
- Assess the effectiveness of environmental management policies in different parts of the Park. Consider your own evaluation, the Warden's views, and the views of visitors (see figure 52).

Resources

Local authorities have different titles for the relevant department. Try *Countryside Management* or *Leisure Services* and *Tourism* to find plans, maps and details for the Country Parks under their jurisdiction. Forest Enterprise have maps and often useful guide books.

A comparison of provision and use of local leisure centres

Starting Points ▶

Leisure centres have increased in number with many urban areas providing more than one. Compare and contrast three centres.

1. The number of people who use each leisure centre declines with distance from each centre.

2. Each leisure centre has a discrete and separate 'catchment area'.

3. Activities at leisure centres have a characteristic pattern of use – daily, weekly, seasonal.

4. Facilities offered at each leisure centre are broadly simliar.

5. Accessibility of leisure centres is similar.

Geographical links to your syllabus

- The distance decay principle can be applied wherever there is movement to a particular place.

- Recreation facilities exhibit a similar hierarchical structure to that in shopping or service functions.

- Increased leisure time and more awareness of personal health has led to increased use of leisure centres.

⚠ *Things to look out for*

This project depends heavily on questionnaires so you must have a very large sample. There are many different areas to cover – age of users, type of activity, time of day. 200 questionnaires in each centre would generate sufficient detail for analysis and interpretation.

Primary Data Collection

1. Collect details of all activities plus costs, frequency offered by each leisure centre.

2. Conduct a questionnaire of leisure centre users to find out what activities they use, how frequently they visit, where they live (use first part of postcode), how they travel, and the reasons for their choice of leisure centre. Record the gender and estimated age of each respondent. Sample different groups of users, using different facilities, at different times of day.

3. Interview manager or trainers to find out how far afield they advertise the leisure facilities. Is there any difference between single events and regular activities?

Secondary Data Collection

1. Leisure centres may have records of earlier summaries of their customers. Most new leisure centres will have a 'target area' within which they wish to attract users.

2. Use OS maps to identify accessibility of locations.

Ideas for recording and analysing data ▶▶

1. Draw a map to show sphere of influence or the furthest distance travelled to each leisure centre. Use one map for all three centres to show overlapping areas.

2. Use a flow line map to show volume of movements.

3. Graph the questionnaire responses – several on one page. Compare similar activities at each centre in terms of volume of users, frequency and cost.

4. Tabulate facilities provided by each centre.

Interpretation and Conclusions

- How do leisure centres differ and why?
- Are they used by people from different areas of a town?
- Is the pattern of use predictable or are there unusual results to explain?
- What factors influence choice of leisure centre?
- Summarise the location of each centre and compare with customer size and location.

Resources

Websites of leisure centres should list all facilities, costs and maps. Look up through the town website.

A study of industrial development in a town

This investigation is suitable for areas which have a number of industrial estates and / or a mixture of old and new industries.

Starting Points

1. There has been a significant change in the type of industries located in an area.

2. Similar industries locate near each other.

3. There is a difference in the location of old and new industrial estates. The older the industrial area the poorer the environmental quality.

4. Are there differences in the location of single plant and multi-plant firms?

5. What factors influence the location of industries?

Geographical links to your syllabus

- Rapid industrial change has characterised most urban areas in Britain.

- Accessibility is a more important locating factor than availability of raw materials.

- Globalisation has reduced the importance of local factors compared with international strategies of location.

- Government or EU influence now plays a significant role in industrial location.

Things to look out for

You may have to sample your industries. Find out what each produces, classify them into groups, then sample from each group.

You must check the availability of information from the industrial units, ie the helpfulness of the managers. They are busy people so the shorter the time for your questionnaire, the better. You may find that some managers are cagey about information. In that case record what you can and select another factory.

Primary Data Collection

1. Identify four areas by their postcodes that contain industrial units. Classify them according to age, size and product.

2. Conduct a questionnaire to the managers of each industrial unit:
   a) Is this a single plant / multi-plant UK / multi-plant international firm?
   b) What does the factory produce?
   c) How long has the factory been in this location?
   d) How much control or influence does this factory have over its precise location?
   e) How many employees does it have?
   f) How important are the factors below in influencing the choice of site for this factory? Answer on a sliding scale from 1 = unimportant, 2 = not very important, 3 = important, 4 = very important:
      - financial incentives
      - local labour force
      - good road communications
      - good rail communications
      - nearby suppliers or customers

Secondary Data Collection

1. Collect a list of factory units from the economic development unit of the local town or city council, county council or local library.

2. Use topographic maps to create topological maps (see figure 54) for index of accessibility (Shimbel Index). Use main roads / dual carriageways / designated ring roads in the town. You can then identify the relative accessibility of each industrial location in the town. Are some locations better than others?

3. Census data will list employment groups for ward and ED areas, and also provide travel to work data. Check the compatibility with postcode areas.

Ideas for recording and analysing data ▶▶

1. Classify industries by type. What percentage of light and heavy industry is there in each postcode area?
2. Map the distribution of industries by age.
3. Apply a nearest neighbour analysis to factories in your different groups.
4. Apply the location quotient to investigate concentration of particular industries in the town.

$$LQ = \dfrac{\dfrac{\text{Number of light industries in an area}}{\text{Total number of industries in that area}}}{\dfrac{\text{Total number of light industries in the the whole study area}}{\text{Total number of industries altogether in the study area}}}$$

5. Draw a star diagram to represent the most important locating factor for factories in the study area.
6. Calculate the Shimbel Index for each industrial area. On your network map include main roads and junctions to other industrial estates as well as junctions leading to motorways or ring roads.
7. Tabulate the relative importance of factors influencing the location of each type or group of factories.
8. Chi Squared test could be used to investigate significant differences between number of industries in your classified groups in each area.

Interpretation and Conclusions

- Has there been a change in the type of industries locating in the town?
- Do similar industries cluster–locate together? Why or why not?
- Do old and new industries show distinct locational patterns? Can you explain those patterns?
- Are all industries influenced by similar locational characteristics?
- How do locational factors differ between single plant firms and multi-plant firms?

Resources

1991 Census
City Council business directory
Resources from the business development unit of local council
Criteria for grants / economic development support from EU / Regional Assistance
Economic development plans / local structure plan – past, current and future strategy website: eurunion.org/infoes/resguide.htm
Commission for New Towns website: www.cnt.org.uk
English Partnerships (for regeneration schemes) website: www.englishpartnerships.co.uk
Welsh Development Agency website: www.wda.co.uk
Business Information Zone website: www.bized.ac.uk/dataserv/datahome.htm
www.bized.ac.uk/stafsup/links.htm

Analysis of traffic flows in a rural area

Starting Points ▶

1. Volume of traffic is related to the number and type of services in a place.
2. The greater the accessibility the more traffic there is.
3. Traffic congestion is associated with short journey distances (ie school run, local shopping).

Geographical links to your syllabus

- Traffic flows vary daily and seasonally.
- Most congestion is caused by local traffic users.
- Higher order settlements experience more traffic, but not necessarily more congestion.

⚠ *Things to look out for*

It's very easy for this topic to fall back into a GCSE-type study.

You must have a genuine question or problem to investigate.

Primary Data Collection

1. Select 10 villages over a fairly large area (400km) with different population sizes. Survey each settlement to identify number and type of functions provided. This will enable you to calculate the centrality index (see figure 47) for each place.
2. In each village make a pilot study to identify which is / are the main routes. Conduct traffic counts at peak morning, mid-morning, weekends during a) school term and b) school holidays. Record the type of transport and number of vehicles. Spend at least 15 minutes on each traffic survey.
3. Identify through traffic by simultaneous traffic counts of vehicles at main routes in and out of the village, and along a main road in the village. Record number and type of traffic.
4. Use car tax discs to record location of licensing office on a sample of parked cars. This will indicate whether the traffic is local, or whether people visit the village.
5. Conduct a questionnaire of drivers leaving car parks to find the reason for their journey and where they live. Usually the first part of their postcode is sufficient, depending on the scale of your investigation.
6. If there is a by-pass, compare traffic flows and type, around and within the settlement.

Secondary Data Collection

1. Investigate local authority traffic survey records. If there are records for all or some of your villages make sure that your primary data will be comparable.
2. Use an OS map to draw a topological map of your area.
3. Find out the population size of each settlement.

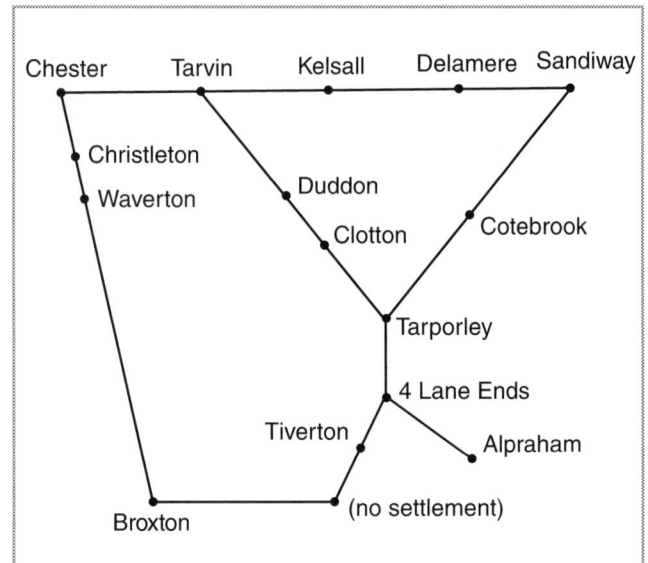

Figure 54 Topological map of a rural area

Ideas for recording and analysing data ▶▶

1. Construct a centrality index (see figure 47) using your survey of services and functions in each village. Do settlements with higher order functions experience the most traffic?
2. Graph traffic data. Use consistent methods throughout with the same scale. Either record all data for each village on one page, or each type of survey for all villages.
3. Draw a flow line map of the whole area to show traffic flows in and out of each village. You could draw a second map to compare flows in school terms with those in school holidays.

4. Conduct Chi Squared analysis to compare differences in traffic a) between villages with similar centrality indices (here you want the results as similar as possible) and b) between school terms and holidays.
5. Draw a topological map to show connectivity of settlements (see figure 54). Calculate Shimbel index to show accessibility of each place.
6. Use a map of the whole area to show residence of drivers who used each village for services (from car tax discs and questionnaire).

Interpretation and Conclusions

- What is the relationship between traffic volume and number of village services?
- How much traffic is generated by shopping?
- Is there a relationship between shoppers' traffic and number of shopping services provided?
- Is there a difference in types and volume and origin of traffic at different times of day / during out-of-school hours?
- Do the larger settlements experience most traffic?
- Do the most accessible villages have the most through traffic? Are they the focus for incoming vehicles? Why?
- Have patterns of traffic flow changed since the local authority surveys were taken? Why?
- If you have identified congestion at particular locations and times, explain why. Think about routes leading in and out of the points of congestion.

Resources

County Highways and Department of Transport have traffic flow data

Investigation of factors affecting agricultural land use

Starting Points ▶

1. To what extent is land use influenced by physical factors such as slope, altitude, soil texture, acidity, stoniness of soil, rootable depth and infiltration rates?
2. Does the intensity of farming decrease with distance from the farm centre?
3. What economic factors affect agricultural land use?
4. What changes have occurred since the 1970s and the entry of Britain to the European Community?

Geographical links to your syllabus

- Several environmental, cultural and economic factors influence the type of farming practised in any particular location. It is rare that any single factor can be identified.

- Type and cost of transport can be very influential.

- Governments may direct the type of agriculture undertaken through subsidies, quotas and initiatives such as the Countryside Stewardship Scheme.

Primary Data Collection

1. Map the current land use of the farm(s).
2. Identify sample sites throughout the farm. A systematic sample using distance from the farm buildings will ensure equal coverage from all fields.
3. At each site measure rootable depth. Use a soil auger to measure soil depth to bedrock. A straightened coat hanger could work, but beware of stones giving shallow readings. Take three different depths to confirm results.
4. Measure infiltration at each site (see figure 23, page 30). Place a tin can or cylinder in the ground. Fill the cylinder with water and record the amount of water required to keep it topped up for two minutes.
5. At each site measure slope angle, soil acidity, organic content and various mineral contents depending on the equipment available, eg nitrogen, potash.
6. To measure stoniness take a quadrat and count the number of stones to a depth of 2cm. Measure the a axis of each stone.
7. With a small soil sample estimate soil texture using the FSC hand chart (see figure 24, page 33).

8. Interview farmer(s): Collect information on farm size, amount of mechanisation, yields from each field, types and amount of fertiliser applied, markets for produce, modes and cost of transport. Questions – how is the land managed year on year? Why farm in this particular way? What subsidies are available? To what extent do they influence farming methods? How important is new technology in the farming system? Do transport facilities influence the farming system? Also ask about human factors, such as what kind of tenure, labour force and farm loans.
9. Ask permission to calculate crop yields yourself. Count the seed heads in a 1m or 0.5m section and find average yield. Measure height of each plant and find average.
10. Look carefully at the land capability chart (see figure 55). Classify each site according to the parameters listed there.

⚠ *Things to look out for*

Make sure you have explained fully to your friendly farmer what you would like to investigate.

If you intend to compare farms, make sure they have at least some features in common. Tell the farmers of your intentions.

Do not disturb livestock or damage crops.

Take measurements with the minimum of disturbance to the soil.

Try to take measurements away from gates, hedges and trees since these may influence soil characteristics.

Secondary Data Collection

1. Draw a geology map of the area. Surface drift geology may be more important than underlying solid geology.
2. Obtain data from Countryside Stewardship Scheme.
3. Collect information about Common Agricultural Policy and specific subsidies or schemes available to the local farmers.

Figure 55 Land capability chart

CLASS	ALTITUDE IN M	WETNESS	SOIL QUALITY	SOIL FERTILITY	SLOPE ANGLE
1 High quality	Below 100m	No limitations Free drainage Rainfall less than 750mm	Deep soil 75cm+ Stone free Loam texture	7+ neutral	Level Not above 3 degrees
2	100–150m	Imperfectly drained Drainage easily modified by liming	Depth 50–75cm Slightly stony	6.0–6.5	Slight Not above 7 degrees
3	150–200m	Some problems but possible to install drainage system	Depth 25–50cm Stony, may have sandy or clayey texture	5.5–6.0	Moderate Not above 11 degrees
4	200–350m	Poorly drained but can be improved to maintain pasture	Shallow – under 25cm Very stony	5.0	Significant 11–20 degrees
5 Poor quality	Above 350m	Poorly drained Drainage almost impossible to install Rainfall over 1250mm	No humus Very stony Skeletal dry soil only	Under 4.5	Steep Over 20 degrees

Figure 56 Example of a summary table of data for each farm

LAND USE ON THE FARM	NITRATE APPLICATION KG/HA	SOIL TYPE	STONE SIZE MM A AXIS	DEPTH OF SOIL CMS
Permanent grassland				
Temporary ley				
Oil seed				
Silage				
Set aside				
Woodland				

Ideas for recording and analysing data ▶▶

1. Construct a base map for the farm(s). Don't forget to use the same scale for each farm. Use this to map land use and topography.
2. Draw isolines of infiltration rates on the base map.
3. Use the choropleth technique on the base map to show stoniness and acidity.
4. Draw a scatter graph of infiltration v slope angle.
5. Graph distance from farm v crop yield. (Use the centre of each field for distance). Use Spearman or Pearson to test the correlation.
6. Establish land use groups, eg temporary grassland, permanent grassland, set aside, different cereal crops if that is appropriate. For each type of land use tabulate

data for stone size, mineral content, soil acidity, soil depth, soil texture and crop yield (see figure 56). Try to summarise each variable for every land use.
7. For each variable you can use Chi Squared to test for significant differences between different land uses.
8. Tabulate the responses to your interviews with farmer(s). Do not be tempted to produce a transcript of your conversation(s). Use the information in an objective way.
9. Map the markets for farm produce.
10. Construct a systems diagram to identify the influences on the land use of the farm (see figure 57). Use arrows of different sizes to suggest the amount of influence.

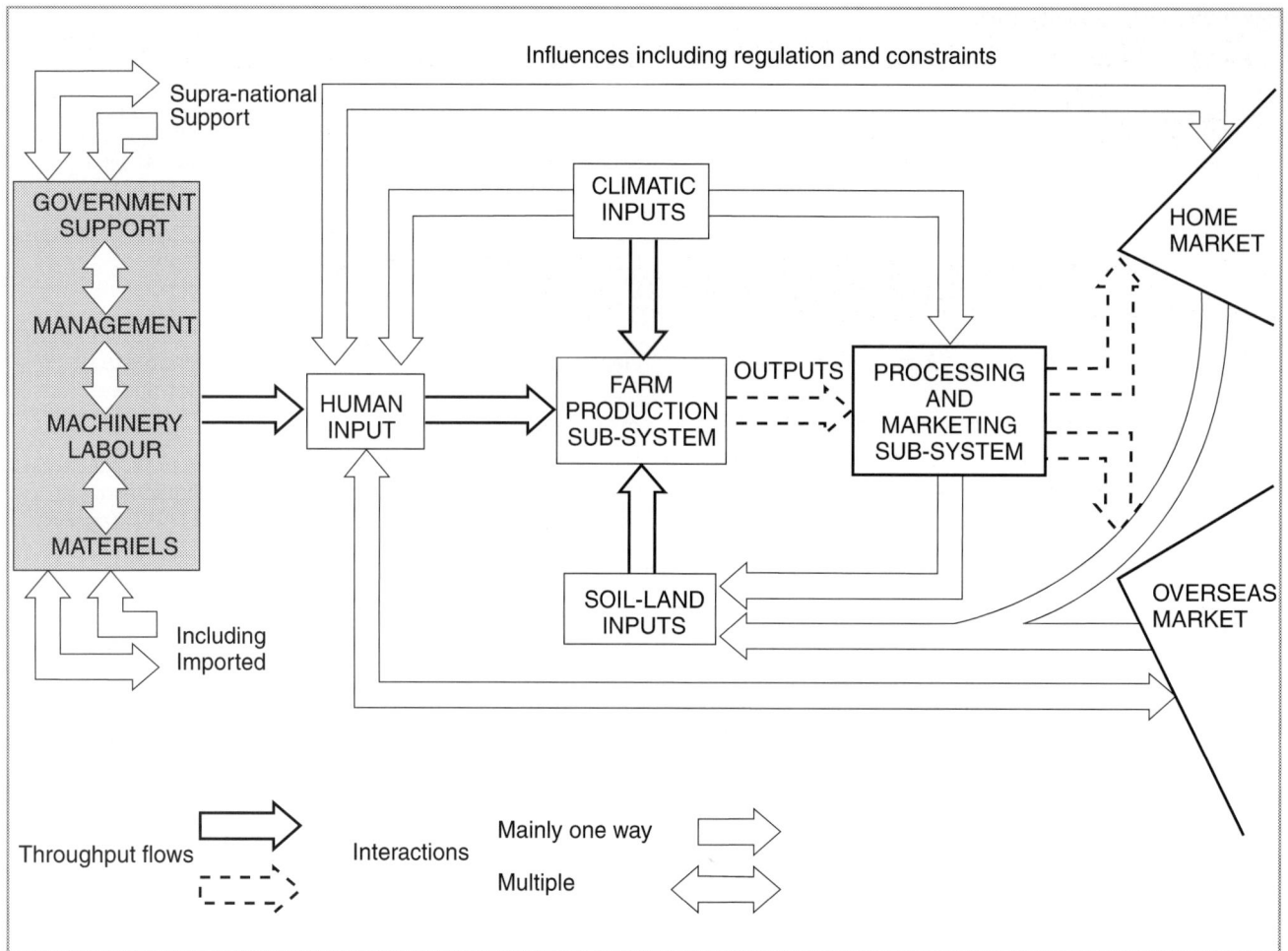

Figure 57 The complex commercial farming system

Interpretation and Conclusions

- Describe the pattern of land use and assess the importance of physical factors in influencing the pattern. Refer to all the variables you have measured.
- Consider any interrelationships there may be between variables, eg soil depth and infiltration and how thay may impact on land use.
- Is there a pattern between crop yield and distance from farm? Why? Make an assessment of the relationship.
- If land use in fields changes over time, what influences those changes?
- How do economic factors explain the pattern of land use? Use information from the farmer to explain the farming system. Avoid simplistic description.
- What are the differences on the farm between 1970 and 2000? Why? Assess the impact of EU policies.
- If you are comparing farms, make sure you do just that. Don't write about one and then the other.

Resources

'Set Aside', Alan Silson, *Geographical Review*, Volume 8, Number 4, March 1995
EU CAP website: www.spfo.unibo.it/spolfo/EULAW/htm
            www.euunion.org/infoes/resguide.htm
MAFF website: www.maff.gov.uk/aboutmaf/regulat/regcont.htm
            www.maff.gov.uk/farm/farmindx.htm
NFU website: www.nfu.org.uk/education
Arable Stewardship Scheme website:
www.maff.gov.uk/environ/envsch.cs.htm
Institute of Arable Crops Research website:
www.iacr.bbsrc.ac.uk/iacr/tiachrome.htm/
Countryside Stewardship Scheme leaflets from Countryside Commision, John Dower House, Crescent Place, Cheltenham, Gloucester GL50 3RA. Tel: (01242) 521381. Fax: (01242) 584270
Ministry of Agriculture, Fisheries and Food Publications, London SE99 7TP
Land use data for parishes from MAFF, Statistics (C and PF branches), Government Buildings, Epsom Road, Surrey TW9 4DU
Parish names can be found in the Municipal Yearbook available in public libraries
National Farmers Union, PA Dept, 164 Shaftesbury Avenue, London WC2H 8HL
Food and Farming Information Service, FFES, National Agricultural Centre, Stoneleigh Park, Warwickshire CV8 8LZ. Tel: (01203) 535707. Fax: (01203) 696388
Food and farming teacher resource centres at:
      Bristol (0117) 923 1800
      Haywards Heath, Sussex (01444) 892700
      Harrogate (01433) 561536
      Halifax (01422) 344555
      Malvern (01684) 892751
      Norwich (01603) 741438
      Lincoln (01522) 522900
      Peterborough (01522) 522900

Analysis of the effectiveness of a park-and-ride scheme

Starting Points ▶

1. There is a significant difference in traffic flows in a town before and after the opening of a park-and-ride scheme.
2. Park-and-ride schemes also affect car parking in the city.
3. There is a significant difference in air pollution before and after the opening of a park-and-ride facility.
4. Park-and-ride schemes encourage visitors from further afield.
5. Park-and-ride buses are principally used by shoppers.

Geographical links to your syllabus

- City centres compete with out-of-town retail centres for trade. Ease of parking is an important factor influencing where people shop. Park-and-ride schemes are therefore important in CBD commercial activity.

- Many city centres cannot cope with the increase in demand for parking.

- Air and noise pollution is a concern in city centres as a result of acute traffic congestion.

⚠ *Things to look out for*

Your personal safety is most important. Do not take risks.

A pilot survey is essential. Sampling for your questionnaire is very important. Consider a systematic sample, ie every nth person. You cannot stratify your sample because you don't know the character of the total population who use the car parks. The proportion in each age group may vary with time during the day. Your pilot survey should give some pointers to your final sampling method.

Figure 58 Index of traffic saturation

This shows how congested the flow of traffic is per hour at any particular location. An ordinary two-lane road has a saturation of 750 vehicles per hour.

Calculate the index of traffic volume, ie the total number of vehicles in one hour.

Index of saturation = 750 – total number of vehicles per hour

Primary Data Collection

1. Before and after opening, count the number of cars along major road(s) affected by a park-and-ride site. Select different times of day to allow for variations between peak congestion and times of low traffic flows. Your pilot study will identify what these times will be. If you wish to identify the number of workers as opposed to shoppers who use a park and ride you will have to record data at 8–8.30am.
2. Question at least 100 people at the bus stop to find out the reason for their visit, how often they visit, where they live (use the first part of their postcode), whether shopping or working in the city, why they use the park and ride, where they parked before the park and ride opened. Note their gender and estimate their age.
3. Collect tax disc information from 100 cars in each of the park and rides and the city centre car park(s).
4. Count the number of cars going further into the city, as well as those turning into the park and ride.
5. Count the occupancy of car parks in the city centre before and after the park and ride opens. Ask at least 100 people where they live, why they use that car park, whether or not they ever use the park and ride. Note genger and age. Note the car park price tariff and that for the park and ride bus service.
6. To investigate the amount of pollution along the main roads into the town centre, attach baby wipes to hedges / lamp-posts at regular intervals. Leave for a week. If this is done before and after a park and ride facility is opened you could compare results. Think about the weather conditions. Dry weather will be more accurate, but the similarity of the weather conditions is more important.
7. Conduct a traffic count of all vehicles along a main two-lane road leading to the town centre (not a dual carriageway). You need to calculate how many vehicles pass by in one hour, so the longer your count time the more accurate you will be.

Secondary Data Collection

The local council will have traffic flow data used in their planning proposals for the park and ride. Make sure your own counts are comparable with these. Local newspapers may have articles and letters commenting about difficulties of parking, solutions and the park-and-ride scheme.

Ideas for recording and analysing data

1. Draw a detailed map of the city area to show the main approach roads and car park locations. Annotate to show the points of congestion and main shopping and office areas.

2. Draw a sphere of influence map of the park and ride using the postcode information. Flow lines from places around the city will indicate the number of visitors. If there are a substantial number of workers rather than shoppers, you could draw different coloured flow lines. Remember that the target area for many park-and-ride schemes is within 6–8km from the city. Repeat this for the city centre car parks.

3. Graph the traffic counts to show 'before' and 'after' volumes.

4. Calculate the index of traffic saturation. (see figure 58).

5. Collect the babywipes you attached to hedges and lamp-posts and grade the 'dirtiness' of each one. You will need to create a scale of white/grey/black and identify by numbers/description (1 = almost black, extremely dirty: 2 = very grey, very dirty: to 5 = little evidence of dirt, clean). Compare the results at different sites and before and after the park and ride opened. Try to draw isolines of pollution levels along the road(s) leading to the city centre.

Interpretation and Conclusions

- Is the park and ride fulfilling its intended function as described by the local council?
- Does the park-and-ride scheme have an impact on traffic and parking in the city?
- What differences are there between the park and ride and city centre car park users?
- Is the sphere of influence of the park and ride restricted to one sector around the city? Why?
- Does the park and ride address the environmental concerns of the local council?

Resources

Park-and-ride advertising
Local Agenda 21
Census 1991 has travel to work data

The social and environmental impact of a by-pass

This study is easier in a small town or village where there are fewer main roads or traffic routes to consider.

There are two approaches to this investigation, which will be dealt with separately:

A – the potential impact of a proposed by-pass
B – the impact of a by-pass after its completion

Starting Points

1. A by-pass is intended to relieve pressure created by through traffic in a town or village.

2. Reduction in through traffic is intended to improve social and environmental quality in the town.

3. By-passes have a significant environmental impact along their route around a town.

Geographical links to your syllabus

- By-pass development involves a balance between costs and benefits.

- Increased concern for the environment has led to concern about the growing volume of traffic in towns.

A – the potential impact of a proposed by-pass

Primary Data Collection

1. Identify the traffic problem in the town or village. Conduct a daily traffic survey at regular intervals during one week, eg at 8.30am, 11am, 1pm, 5.30pm, 8pm. Note the type and volume of vehicles and, if possible, the number of passengers. Select sites on the main roads that will have a direct link to the proposed by-pass. Remember to coordinate your sampling sites with traffic surveys completed by the local council. Consider evaluating traffic congestion in the town centre if your settlement is quite large. Take photographs to support your results.

2. Measure noise levels during each traffic count and in streets 300–500m from the main road.

3. Estimate pollution levels using baby wipes around lamp-posts at each study site. Leave for a week and compare the accumulation of dirt at each location.

4. Conduct a questionnaire of people living, working and walking along the main routes currently congested, and also those people directly affected by the new building, to assess local views about the proposal. Use a stratified sample to cover residents, shopkeepers, pedestrians.

Questions could include:
- where do you live (first part of postcode)?
- are you a resident / shopkeeper / pedestrian using the route?
- how do you rate the current traffic situation in terms of noise, air pollution, safety, effect on local trade? Answer on sliding scale from 1 = poor, 3 = satisfactory, 5 = good.
- do you think the by-pass is necessary?
- how will it affect you personally?

5. Draw a detailed map of the proposed by-pass route. Identify the land use which will be affected. Try to count or measure the number of trees and hedgerows that will be affected. Conduct an environmental impact assessment (see figure 59) along the planned route. If the by-pass is long, you may decide to evaluate its impact at several points along the route. Take photographs to illustrate the EIA you make.

6. Interview local farmers to asses their views on the proposal.

⚠ *Things to look out for*

Your personal safety is most important. Do not take risks with traffic.

Integrate your sampling sites with your secondary data sources.

Think about the timing of your traffic counts – school holidays are significant.

This can be an emotive subject. If you are interviewing people, state clearly and simply what you are doing and why. Do not take a personal view until you have completed all your analysis.

Figure 59 Environmental impact assessment of the building of a by-pass

CHARACTERISTICS OF EXISTING ENVIRONMENT	WEIGHTING	CONSTRUCTIONAL PHASE				OPERATIONAL PHASE
		LAND REQUIREMENT	LABOUR REQUIREMENT	BUILDING WORK	TRANSPORT AND MATERIALS	VEHICLES USING ROAD
Landscape	9	M 27		M	S 9	I 18
Floral fauna	8	M 27		M	S 8	I 16
Employment	4	–	M	S		
Housing	4	S	S			
Traffic flow	3		S	S	I	M
Air pollution	7			I	I	M
Noise pollution	7			I	I	M
TOTAL SCORE						

Impacts assessed as the following: Major impact,   M = 3
Some impact,   I = 2
Slight impact,   S = 1

Multiply the impact by the weighting for each characteristic.
Total score from each column indicates the areas of greatest impact.

Figure 60 Environmental quality assessment of the by-pass proposal

This can be assessed in two places – the by-pass itself, and the main roads which will be affected by the new road. These are ideas to use, as well as the planners' criteria. Compare your results with those presented as part of the planning application. Try to be as objective as you can. Consider asking other people for their reactions.

	Significant better environment	No change	Significantly worse environment
Score	5	3	1

Noise
Air pollution
Vibration
Smell
Landscape quality
Traffic congestion
Public safety

The lower the score, the less beneficial the by-pass would seem to be.

Figure 61 Ideas for the cost benefit exercise

	COSTS	BENEFITS
Social	Disruption to local residents during construction	Reduce accidents Work for local area during construction Safer environment for residents
Economic	Costs to taxpayer and local business Loss of farmland	More customers attracted to local shops because improved environment
Environmental	Air pollution and noise during construction	Less traffic therefore less noise and air pollution

Secondary Data Collection

1. Obtain detailed planning proposal for the by-pass, including local traffic surveys.
2. Find out the values for land on which building is planned, and for land nearby which will be unaffected by the new road (as a comparison). Try to find productivity figures for the land which it is intended to build on.
3. With reference to the planning proposal document, make an estimated environmental quality assessment (see figure 60) of the scheme. Consider carefully the comments in the proposal concerning traffic noise levels, any environmental protection measures such as cuttings or woodland screening proposed by the Highways Department.
4. What evidence does the local council have of environmental stress along existing roads – collapsed sewers, need for resurfacing of roads, number of accidents, road obstructions, pressure on parking? Can you support this evidence with your own observations?
5. Collect evidence on the costs of the by-pass scheme including compensation to be paid to landowners.

Ideas for recording and analysing data ▶▶

1. Draw a detailed, annotated land use map of the by-pass proposal and the roads to be affected.
2. Draw flow line maps of traffic. Use different colours for different types of traffic, eg bikes / cars / commercial. Incorporate passenger data if you measured it. Use annotated photographs to show the extent of the traffic.
3. Graph the noise and pollution levels and compare the main road and more distant sampling sites. If you used baby wipes generate a scale based on depth and extent of dirt or pollution.
4. Graph the questionnaire responses, several per page, to compare the views of people along the main affected roads and those affected by the by-pass route.
5. Summarise the results of the environmental quality assessment and the environmental impact assessment. Annotate your phototgraphs carefully.
6. Tabulate a cost benefit summary of the by-pass proposal (see figure 61).

Interpretation and Conclusions

- Draw together all the evidence for traffic pressure on roads leading into the town and comment on the need for the by-pass. Is there a congestion problem? Where? When? Why?
- Using the cost benefit analysis and the EIA discuss whether the proposed route is appropriate, and whether or not you think it will solve the traffic problems of the area. Are the benefits worth the costs? How do your results compare with the argument put forward by the planners?
- Has the best route been chosen, or are there alternatives which could be explored further?

Resources

Local structure plan from local council
Details of the by-pass planning proposal submitted to the local authority
Highways Agency – National Trunk Roads Statistics Department
County Highways department
Look for websites established by opponents to the scheme
Local newspaper reports
Attend a meeting held as part of the local council's consultation exercise
Environmental Impact Assessment (refers to a proposed by-pass)
Geography Review, Volume 7, Number 5, May 1994

B – the impact of a by-pass after its completion

Primary Data Collection

1. Count traffic amount and type along the by-pass each day at different times.
2. Similarly, count the traffic along the most likely alternative routes to the by-pass.
3. Conduct an environmental quality assessment (see figures 60 and 63) of the by-pass area and the former congested routes to assess the effectiveness of the changes to traffic flow. Use photographs to support your results. You will need to make an assessment in several locations. Alternatively, you could ask several local residents to complete an EQA for before and after the road was built.
4. Conduct a questionnaire of local residents who lived in the area before the by-pass was opened to find out what they think of the development now that it has happened (see figure 62).
5. Estimate the travel times between one side of the town and another. Drive between two points at similar times to test the journey time. Has the by-pass saved time?
6. Draw a current land use map to show new land uses since the completion of the road.
7. In local car parks, use tax discs to identify where people live (ie assume they tax their cars near their home). Are they local people or visitors from further afield?

Figure 62 Questionnaire to local people after by-pass has been built

1. Postcode
2. Were your original views on the building of the by-pass positive or negative?
3. Do your views differ now? Y / N
4. How often do you shop in this town? Daily / Every 2–3 days / Weekly / Occasionally / Rarely
5. As a result of a decrease in traffic, do you think the town has become an easier and more pleasant place to shop? Y / N
6. Has the building of the by-pass affected the number of times you shop here? Now shop more / Now shop less
   Where do you shop now? Why?
7. Have you used the by-pass frequently since its completion? Y / N
8. If travelling from one side of the town to another, would you find it quicker to use the by-pass or go through the town? By-pass / Town route
9. Do you use the by-pass to go to the nearest town?
10. As a resident, has the building of the by-pass had any effect on the area in which you live? If so, how?

Secondary Data Collection

1. Collect all the data from the initial planning proposal and traffic surveys prior to building the by-pass. There should be information on noise, traffic volume and type, on local roads, as well as reports of the public consultation meetings. Local councils have to report on their monitoring of major traffic schemes before and after completion.
2. Research house and land values for areas affected by the by-pass. Try to compare current values with those before the road was even planned. Are there any contrasts evident?
3. Draw a land use map to show former land uses in the area taken by the by-pass.
4. Find the final costs of the by-pass scheme and compare with those estimated in the planning proposal.
5. Classify local press reports (see figure 65) to compare local views before and after the scheme was open.

Figure 63 Environmental quality survey

You may be able to ask people who have lived in the area for some time to complete this EQS for before and after the road was built.

Use a sliding scale such as:
1 = very poor, 2 = poor, 3 = satisfactory,
4 = improved, 5 = much improved

Consider factors such as noise, dust, smell, vibrations, personal safety and attractiveness of area.

⚠️ *Things to look out for*

Beware biased views of local people and planners. Try to be objective.

Ideas for recording and analysing data ▶▶

1. Draw flow line maps of traffic at different times of day and week. Incorporate types of traffic too. Compare with traffic flows before and immediately after the by-pass was opened. Apply Chi Squared to test for significant difference between traffic volume before and after the road was built.
2. Graph the EQA responses and compare with similar secondary data if it is available. Annotate photographs of different locations to justify your assessment.
3. Graph questionnaire responses. Organise several graphs on one page so that you can compare easily.
4. Draw a sphere of influence map using car park data and comment on the distribution shown.
5. Tabulate local press reports. Have views on the by-pass changed? Why?

Interpretation and Conclusions

*   Has the by-pass achieved its desired effect as identified in the planning proposal?
*   Comment on the costs and benefits of the scheme in the light of the estimated and actual financial costs as well as the social and environmental costs.

Resources

Planning documents and reports of consultation meetings
Local press reports
Old land use / OS maps of the area
Postal Address Book
'Transport – Projects towards sustainability: 1. Local transport fieldwork', David Job, *Geography Review*, Volume 9, Number 1, September 1995
Transport 2000: Walkden House, 10 Melton Streeet, London NW1 2EJ. Tel: (0171) 388 8386
Sustrans: 35 Kingfisher Street, Bristol BS1 4DZ. Tel: (0117) 926 8893

An investigation into the impact of a proposed new sports centre on the local community

Starting Points ▶

1. Why is the new sports centre being proposed? Is there a demand for new facilities?
2. How will the new development affect local public and private traffic flows?
3. How will it change environmental quality and land values?
4. What do local residents feel about the proposal? Who is intended to use the new facility?

Geographical links to your syllabus

- Growth in the leisure service industries provides employment opportunities in the tertiary sector.

- New sports complexes are frequently part of redevelopment schemes.

- Leisure can be expensive and there are socio-economic patterns to investigate.

- There is a hierarchy of leisure facilities with different sizes of sphere of influence. How do new centres fit into the existing framework?

⚠ Things to look out for

Traffic survey – make your data comparable with council or developers' data.

How many sites give a good picture of impact? When? Where? For how long? Traffic flows vary considerably during the day and week.

Sampling of your questionnaire. How many people? How will you choose? A stratified sample will give views from different ages and social groups of people, and that could be significant.

Which other sports centres? Explain your purpose and ask permission.

Get the support of the developer at the outset. They will have lots of information.

Primary Data Collection

1. Conduct traffic surveys along roads which could be affected by the development. Consider time of day and types of traffic.
2. Conduct an environmental quality assessment (see figure 64) at the proposed site and the surrounding area. Choose sites at varying distances from the development – the number depends on statistical technique.
3. Make annotated sketches or photographs of the site.
4. Collect current land values for the area of the development from the local council or Inland Revenue.
5. Gather the responses of local residents. Complete a questionnaire survey to identify residents' views on the impact of the development – noise, environmental quality, land values. Do they support the scheme? Would they use it?
6. Consider the impact on other facilities in the area. What facilities do nearby sports centres provide?
7. Conduct a questionnaire of users at nearby centres to establish sphere of influence. Will there be overlap with existing centres? Will people change the venue of their activities? If so, what effect will that have on existing sports centres?

Secondary Data Collection

1. Obtain OS maps to analyse location.
2. Collect details from the developer regarding location, plans, facilities, anticipated traffic flows. Are there new services to be provided by local bus companies?
3. Make an environmental quality assessment from the plans / proposals provided by the developer.
4. Analyse local press reports and letters to local newspaper (see figure 65).

Figure 64 Environmental quality assessment score sheet

	GOOD				POOR
Category	+2	+1	0	−1	−2
Vandalism					
Noise level					
Air pollution					
Open space					
Landscaping / visual quality					

Remember that you should define what you mean by each category, and what each score represents. The higher quality characteristics earn the highest score. For instance, a high value for landscaping could mean mature trees and bushes, well-kept grass spaces and attractive buildings, while low score would represent almost total lack of trees or grass, unkempt appearance, drab buildings.

Figure 65 Analysis of press reports

TOPIC COVERED IN REPORT	TRAFFIC CONCERNS	APPEARANCE	DISRUPTION	LOCAL NEED / COMPETITION	COSTS (USER AND RATEPAYER)
Supporters of the scheme					
Opponents of the scheme					

Ideas for recording and analysing data ▶▶

1. Use annotated sketch maps or overlays to show change in land use.
2. Use flow line maps and annotated graphs to compare primary and secondary transport data. Bar charts should be clearly associated with specific map locations.
3. Use isolines on a base map to show environmental quality assessment results. Look for relationships between sets of data, eg noise levels and EQA. Is there a difference between the environmental quality before and after the development?
4. Draw a choropleth map of land values. This might be extended to create an anticipated land value map by using the EQA results. Are particular areas likely to be affected more than others? Why?
5. Draw a variety of bar and pie charts to represent views from questionnaires. Beware of overkill. Place several on one page so that you can identify relationships between your responses. Consider carefully which answers are related to each other, and where you can make connections between the answers.
6. Map the sphere of influence of each sports centre you included to show whether or not there is any overlap of market area. Don't forget to annotate to indicate reasons for their shape – road network, topography, social class of particular areas.

Interpretation and Conclusions

- Refer back to your original statements and comment on each one.
- Summarise the potential impact of the new development in the following ways:
  a) effect on transport
  b) impact on the environment
  c) impact on local people
  d) impact on other sports centres.
- Give reasons for your conclusions. You may only be able to *suggest* reasons – that's fine because you can then develop ideas for further investigation.
- For every conclusion you reach, clearly state the evidence and reference the relevant graph, map or statistics.
- You may find that the evidence supports the planners' views, but don't be afraid to contradict the views of the developers if that is what your evidence suggests.

Resources

Documents from planners and developers concerning similar new developments elsewhere, particularly those built by the same company
Guidelines from the Department for the Environment / EU regarding planning proposals
Local structure plan

The impact of a new supermarket on the local environment

The impact of a new supermarket on the local environment

Starting Points ▶

1. New supermarkets are often built on greenfield sites on the edge of towns with a consequent impact on the local environment.

2. Supermarkets may have a detrimental effect on small local retailers.

3. There is a significant change in traffic flows as a result of new supermarket developments.

4. New supermarkets often attract other chain stores to form a cluster of out-of-town retailing.

Geographical links to your syllabus

- Shopping patterns have changed with the advent of out-of-town retail developments.

- There is concern about the urban sprawl encouraged by such developments.

- New shopping developments affect the transport network.

Primary Data Collection

This topic can be studied by focusing on the planning application alone. However, the effect of building a supermarket can be estimated by comparison with a similarly located supermarket recently opened. This will help to support the element of prediction involved in working on a planning proposal. While no two developments are identical it is possible to estimate traffic flows and the impact of the supermarket on established retailers by careful comparison with the impact which occurred elsewhere.

1. Map the current land use of the proposed site and local food retailers.

2. Conduct a questionnaire of 100 shoppers in the local area to find their attitudes to the new supermarket proposal. Is the development needed? Would they shop there? Why or why not? Would they still use the local retailers? How frequently would they shop in either location? How would they travel?

3. Find another similarly located supermarket for comparison and question 100 shoppers there. Where do they live? How often do they shop? How do they travel? Why do they use that retailer? Do they use small local retailers? Where did they shop before that supermarket was open? Do they combine a food shopping trip with other shopping needs?

4. Investigate noise levels at the proposed supermarket site and compare with noise levels at the existing supermarket.

5. Observe the proposed supermarket site carefully and estimate the potential loss of hedgerows and trees.

6. Conduct a traffic count on existing roads. Think about variations during each day and week. Collect similar traffic data at sites near the established supermarket.

7. Question the local shopkeepers to find out how their trade or turnover was affected by the established supermarket. Question local shopkeepers who could be affected by the new supermarket to find their views on the new development. Ask different groups of retailers in different parts of the town. Remember that their views will be biased in the face of potential competition.

8. Make an environmental impact assessment of the proposed development (see figures 59 and 66). You should adapt the environmental characteristics to suit your particular situation.

⚠ *Things to look out for*

Remember that this can be an emotive subject.

Questionnaire respondents will have a particular view for or against the proposed supermarket and this will colour their answers.

Justify your sample size and sampling technique. Remember that 100 questionnaires is only a tiny fraction of the customer population.

You will need permission to undertake a questionnaire near some supermarkets.

Secondary Data Collection

1. Collect full details of the planning application, including artist's impressions of the development. Conduct a bi-polar analysis on the area. How do the two compare? There will be extensive displays in local libraries and council offices.

2. Use the Census to establish population size and the local structure plan to find out predicted population growth and land use zoning of futher developments of housing and retailing in the area.

3. Collect traffic flow data for the affected roads from the local highways office.

4. Collect the details of the planning proposals for the existing supermarket.

Physical, chemical, flora + fauna, cultural factors

Figure 66 Environmental impact assessment data collection table

TYPES OF CHANGES THAT MIGHT HAPPEN	CHARACTERISTICS OF THE ENVIRONMENT THAT MIGHT BE AFFECTED										
	Soils	Surface water	Trees	Air quality	Animals	Land use	Employment	Population	density	Noise	Total
Modification of habitat							–				
Urbanisation		4/3									
Landscaping							–				
Modification of ground cover											
Roads and footpaths									3/4		
Traffic											
Surface drainage		3/4									
Pollution											
Noise and vibration	–										
Total											

Total score for each section

The size of the project's impact (on a scale of 1–4)

The importance of the impact on the characteristic (on a scale of 1–4)

A diagonal line through a box shows that an impact will occur

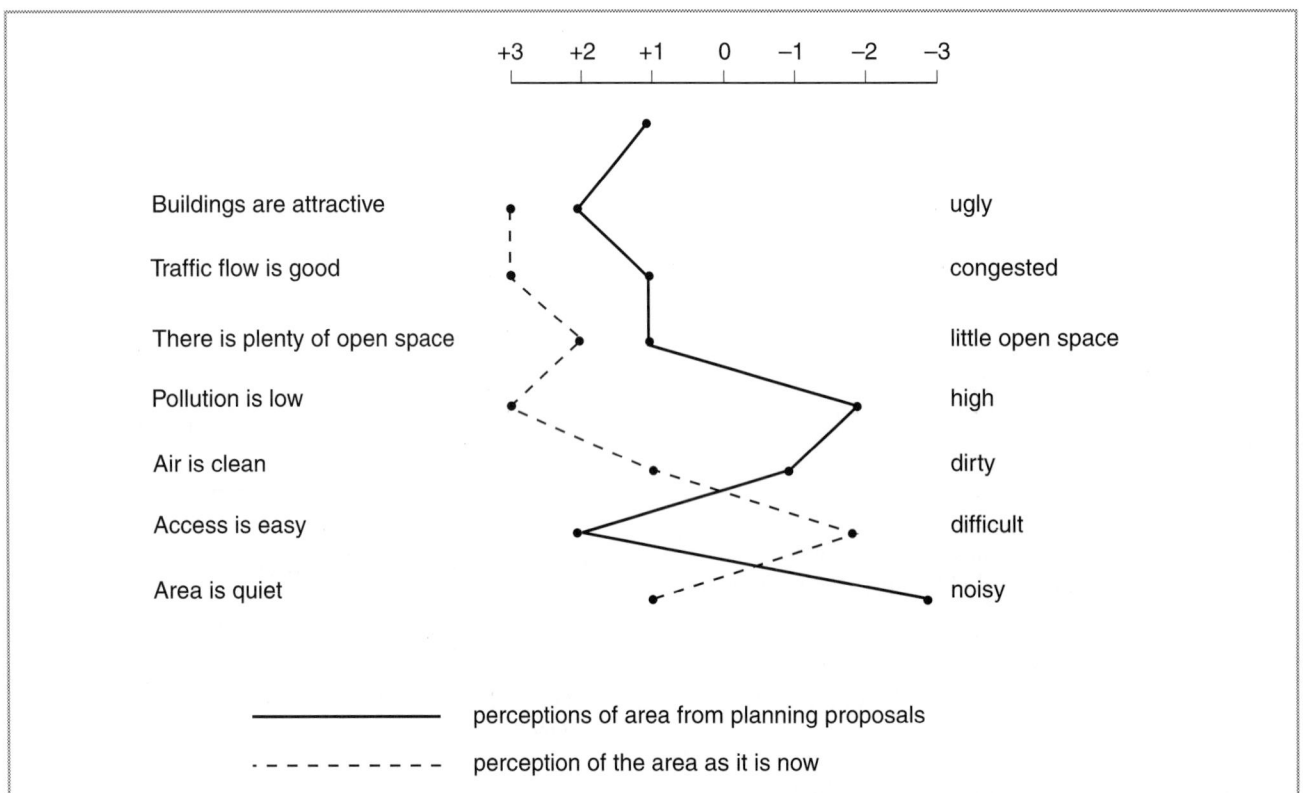

Figure 67 Bi-polar semantic scale to quantify perception

Ideas for recording and analysing data ▶▶

1. Draw annotated maps to show the existing land use and proposed development.
2. Graph the questionnaire results to show attitudes to the new development, and compare with similar responses from the existing supermarket.
3. Draw flow line maps of traffic data using your own counts and local council data. Compare with predicted patterns given in the planning proposal. Similarly compare the predicted and actual traffic flows for the existing supermarket. Were the planners right in their predictions for the existing store? What does that suggest about the estimates for the new store?
4. Use the bi-polar diagram (see figure 67) to compare the predicted appearance of the environment in the planning proposal with the current site for development. Compare the number of trees and hedges which will disappear with the landscaping intended by the developers. Look for evidence of storm drains, new ditches, the amount of tarmac or impermeable surface compared with the current situation.
5. Tabulate the EIA results and comment on the areas of greatest impact. How are they being addressed in the planning proposal?

Interpretation and Conclusions

- Summarise the impact that the proposed supermarket will have on the environment. Remember to refer to the structure plan to incorporate the council's long-term proposals. Does it seem that one development will lead to others?
- How will shoppers and retailers be affected by the new store? Will traffic flow become a problem?
- Does the evidence from other similar supermarket developments suggest that the planners have made realistic predictions?
- Summarise the proposal with a cost benefit evaluation (see figure 61, page 84). Is the new supermarket needed?

Resources

Local structure plan
Detailed planning proposal
Planning documents for another similar supermarket development
Look for websites created by local groups, often opponents, affected by the development

The impact of a new housing estate on the surrounding area

Starting Points ▶

1. How has the area changed since the development began?
2. How do local people view the new estate?
3. To what extent does the new estate affect the natural environment?
4. What has been the social and economic impact of the new estate?

Geographical links to your syllabus

- Although increasing numbers of people live in urban areas there is also substantial counter-urbanisation as people try to avoid areas of poor environmental quality in the centres of towns.

- The changing structure of the population is creating a huge demand for new homes in Britain.

⚠ *Things to look out for*

Be aware that a new housing estate can be a very sensitive issue.

Justify your sampling technique and sample size.

Primary Data Collection

1. Map the land use of the area.
2. Conduct a landscape evaluation of the estate (see figures 44, page 58 and 70, page 95). Use photographs to support your observations.
3. Interview the landowner who sold the site. Find out how productive the land was.
4. Summarise house prices in the area and the types of houses built on the estate.
5. Conduct a questionnaire of 100 local residents (not living on the estate) and local shopkeepers (see figure 68).
6. Interview the many groups concerned with the estate such as the parish council, residents' association and political parties to find their perception of the impact of the new estate (see figure 68).
7. Interview a local head teacher to find out the impact on school numbers and other social impacts.
8. Survey the traffic at a point where it enters and leaves the new estate. Note the numbers leaving and entering at different times of day and week.
9. Sample 100 cars in the estate and note the registration letter. Record the number of caravans and second cars (visitor parking is a problem here, but choose a time such as 6–7pm when many people are likely to be at home). This will give an indication of the economic status of residents.

Figure 68 Housing estate questionnaire

Do you agree or disagree with the following statements? Use a sliding scale from 1 = strongly agree to 5 = strongly disagree.

1. The housing estate was a much better option than an open field and more space.
2. The extra vehicles cause little extra pollution.
3. There is very little chance of an accident where the roads meet.
4. The new residents cause few social problems.
5. The local economy and schools benefit greatly from the extra residents.
6. A housing development on this site is long overdue.
7. The new estate is a good development.

Secondary Data Collection

1. Use aerial photographs and large-scale maps in the local library to map land use before the estate was built.
2. Classify and summarise local press reports concerning the new estate (see figure 65).
3. Look at leaflets from the estate builders to find out what sort of homes they intended to build. Detailed plans should show you what modifications were necessary to drainage and vegetation.

Figure 69 Analysis of questionnaire responses

Number of persons answering questionnaire

The total score for each question.
The higher the total score for each question, the more disagreement there is of the statement

PERSON NUMBER	1	2	3	4	5	6	7	8	9	10	11	12	TOTAL
Question 1	4												
2	5												
3	3												
4	2												
5	4												
6	3												
7	4												

Example of recording sheet for questionnaire
MAX SCORE 7 × 5 = 35

Sample response from person 1. A score of 25/35 shows that the person who filled in the questionnaire was fairly strongly against the new estate.

Ideas for recording and analysing data ▶▶

1. Make an annotated map of the estate with overlays, identifying major land use changes. Try to quantify the types of land use changes. Comment on changes to drainage patterns, numbers of trees and length of hedges.
2. Graph the questionnaire results, several on one page for comparison (see figure 69). Tabulate the comments from the local groups and the head teacher.
3. Annotate photographs of the estate to support your landscape evaluation. Use evidence of car ownership and house prices to comment on the social characteristics of the estate.
4. Draw flow line maps of traffic data. At what time of day is it busiest? Is the traffic mainly commuter-generated?
5. Summarise newspaper reports about the estate. Have local views been justified?
6. Make a cost benefit analysis of the development of the estate.

Interpretation and Conclusions

- How has the area changed since the development began?
- How do established residents feel about the new estate?
- Have social and economic impacts outweighed changes to the natural environment? Use your information from the landowner to assess whether the development was worthwhile.
- Is the profile of the new development matching that predicted / expected by the local council?

Resources

Local structure plan
Planning application details for the estate
Leaflets from local builder

Investigation into the impact of a golf course development on the physical and economic environment

Starting Points ▶

1. Golf courses may provide environmental protection from the expansion of the built-up area.
2. Golf courses support the local community by providing employment.
3. The benefits of golf course developments outweigh the costs.

Geographical links to your syllabus

- Recreation has a significant impact on the physical environment.
- Leisure facilities are important contributors to the local economy.

⚠ *Things to look out for*

You must not disturb the golf course land in any way.

Make sure your questionnaires are very focused. You need a good deal of information and you must not waste anyone's time.

Become informed about the golf course and management practices in general before you interview the greenkeeper. They will be much more likely to help you if you are well informed.

Primary Data Collection

You will need a good contact in a golf club to make this study work well.

1. Collect data for the number of people employed full time and part time at the club as groundspeople and in the clubhouse. What percentage live within 5–20km radius? Identify the age structure of the workforce.
2. Interview the golf course manager or head greenkeeper to collect details of the physical environment of the course. How long has the course been open? What sown grasses are used in different parts of the course? Which fertilisers and top dressings are used, when and how much? What machinery is used on different parts of the course? What functions do the machines perform and why? How much irrigation is required on the course,

when and why? What birds and mammals are regularly sighted on the golf course? Are there nesting sites?
3. Make a large-scale, detailed land use map of the golf course on graph paper to calculate the proportions of greens, rough, tee / fairway, water and woodland.
4. Make a woodland survey of types and number of trees and hedges on the course.
5. Use a stratified sampling method to measure soil compaction at different points on the course including footpaths. Use a soil penetrometer, or try a worm count. (This could be difficult since groundspeople may not allow you to disturb the greens.)
6. Take photographs of the golf course environment and compare with photographs of land over the boundary. Identify the natural physical features in each landscape. Make an index of visual quality for golf course land and adjacent land (see figure 70).
7. Measure the chemical content (such as pH, nitrates and phosphates) of streams flowing into the golf course area. Repeat these measurements on streams flowing out of the golf course area.
8. Make a thorough survey of club members during a busy weekend. (Beware of particular competition days.) Near the club entrance count the number of members using the course, note their gender and estimate their age. Record the registration letter of cars and the place of registration.
9. Question 20 local residents or farmers surrounding the golf course. Do they approve of the golf course? Do they think it protects the environment? Is there traffic congestion on local roads? Are there any ways in which the golf course is regarded as detrimental to the environment? What amount of fertilisers or chemicals do farmers add to their land and when?

Secondary Data Collection

1. Find out the costs of playing at the club and the length of waiting lists. What does this suggest about the socio-economic status of members?
2. Collect data for agricultural land in the area from the county council research and intelligence unit. Establish whether the golf club generates more employment per hectare than farmland. Look for details of number of farms and hectares of farmland to calculate average area

per farm. You should be able to get details of labour – farmers, managers, regular workers and seasonal workers. Calculate the average number of people per farm and then the hectares per employee (see figure 71).

3. Refer to old OS maps of the area to establish land use before the golf course was built.

Figure 70 Index of visual quality

SCORE	+3	+2	+1	0	−1	−2	−3
Natural features							
Change of gradient							
Natural woodland							
Lakes / ponds							
Natural colour							
Wildlife							
Noise							
Smell							
Residential features							
Services / shops							
Communications							
Noise							
Smell							
Afforestation							
New hedgerows							

Add all the figures to reach a score for each site, then divide this by the number of features you have used to measure impact.

A maximum score of 3 shows that the facility has greatly enhanced the quality of the local environment. Similarly a score of −3 suggests that the facility spoils the environment.

Figure 71 Statistical data for Cheshire

	1933	1994	1995
Land (hectares)	168,502	167,486	166,038
Farms, Number of holdings	4,450	4,422	4,180
Average farm area (hectares)	37.87	37.87	39.72
Manpower			
Farmers	6,390	6,354	6,153
Managers	145	161	139
Regular workers	3,422	3,445	3,382
Seasonal workers	1,333	1,326	1,486
Total	11,290	11,286	11,160
Average manpower per farm	2.54	2.55	2.66
Hectare per employee	14.91	14.85	14.93

Ideas for recording and analysing data ▶▶

1. Make a detailed, annotated map of the golf course to show different land use categories. Integrate photographs to illustrate the types of areas over the course. Compare this map with a map of the area before the course was built. Summarise how the land use has changed.
2. Collate a series of graphs to show data for golf course members – age, socio-economic status as shown by age of cars, where members live. Draw a map to show the sphere of influence of the golf course.
3. Compare the employment per hectare created by the golf course with that provided on farmland in the area. Tabulate the employment opportunities offered by the golf course.
4. Summarise the course maintenance undertaken by the grounds staff.
5. Compare the chemical content of streams flowing in and out of the golf course area.
6. Tabulate data for compaction on different parts of the golf course. Use a one tailed Chi Squared test to see if there is a significant difference between compaction on different parts of the course. If there is such evidence discuss how and why such compaction occurs and assess the effect it has on the physical environment.
7. Tabulate / graph details of wildlife sighted on the golf course. Also the types of trees and number of each species.

Interpretation and Conclusions

- Discuss the extent to which the golf course protects and harms the environment.
- How is the golf course managed in order to sustain the natural environment?
- What are the socio-economic characteristics of the golf course membership? Are members local residents? Is the workforce local?
- Compare the numbers of people employed by the golf club with those for the local farming community.
- Make a cost benefit analysis of the impact of the golf course on the local community (see figure 61).

Resources

Golf course website, plus others in the area for comparison
County statistics to agricultural land and employees

The impact of a professional football club on the local environment

Starting Points ▶

1. Football clubs can be seen as both a benefit and a burden on the local community.
2. As distance from the football ground increases, its impact decreases.
3. The costs and benefits are viewed differently by different groups of people.

Geographical links to your syllabus

- Leisure activities, such as football, have a significant impact on the economic and social life of local communities.
- Large leisure facilities, such as major football clubs, may be part of urban redevelopment schemes.

Primary Data Collection

1. Make a detailed land use map of an area approximately 1km radius around the football ground.
2. Identify at least 10 sampling points on a systematic basis.
3. At each point record land use and housing density – count the number of front doors along a 100m section of pavement to give an indication of density.
4. Count the number of pieces of graffiti in a 100m section of street. Note how many are football related.
5. On a non-match day at each site count the number of parked cars along the section of street, and the number of pieces of litter. Repeat this on the morning of a match day, while the match is in progress, and also after the match has finished.
6. Measure noise levels on a non-match day with those on a match day, during and shortly after a match.
7. Identify the main road routes to the ground. Count the amount, type and direction of traffic along these routes on a non-match day, and before, during and after a match. It is likely that traffic intensity will be greatest after the game.
8. Survey all the traders located around the ground – newsagents, hot dog stands, newspaper sellers, car park attendants, take-aways, pubs etc., to find what proportion of their trade comes from football fans on match days.
9. Conduct a questionnaire of local residents to find out how they view the football ground. Do they support that particular club? Are they disturbed by vandalism / noise? Is the car parking by fans a problem? Do they benefit in any way from the football club? Does the ground affect house prices in the area? Justify your sampling procedure and record age and gender of each respondent.

10. Use local estate agents to find out house prices at each site over your study area. Try to compare similar types of houses, although this may be difficult.
11. Interview the Community Development Officer of the football club to find out what community schemes are supported by the club. Try to find out about full- and part-time employees, especially for match days – what proportion live locally? How many people are employed by the club? If you have identified the key club-related issues which concern local residents you may be able to find out what is the club reaction / response.
12. You may decide to compare the responses of local people to those of people living some distance away from the football ground (see John Bale's article in *Geography* listed below) and try to establish whether the problem of football grounds is a real or perceived one.

⚠ *Things to look out for*

Do not assume that the football club will give you much information. Not only are they busy people, they have many requests for data, and do not always have the sort of records you would expect from large organisations.

Even if data is promised, do not rely on it.

Secondary Data Collection

1. If the football ground is in an area of urban redevelopment or regeneration look for the redevelopment or planning proposals.
2. There may be local council traffic data for the areas of congestion around the ground.

Ideas for recording and analysing data

1. Use photographs incorporated with a map to descibe the area surrounding the football ground.
2. Make an outline base map centred on the football ground on which you can draw isoline maps using data collected at different sites for noise, amounts of graffiti, litter, house prices or the results of your questionnaire. Can you identify any similar patterns between these maps? Are there any relationships which would be worth investigating further?.
3. Draw scatter graphs to show changes in these variables over distance. If there is a linear relationship test this using correlation technique such as Spearman Rank. To compare the patterns at different times such as between match and non-match days use the Mann-Whitney U test. Don't forget to test your results for significance.
4. Draw flow line maps to show traffic patterns at different times. Make the maps small enough to fit several on one page so that you can compare patterns more easily. What are the differences and why? Comment on which road junctions are most affected and why.
5. Analyse the economic impact of the football ground by graphing the responses from local traders. Assess the extent to which their trade is affected by football.
6. Similarly, graph the questionnaire responses, by age and gender. Annotate the graphs to identify the key issues which concern local residents.

Interpretation and Conclusions

- How and to what extent does the football club affect the physical environment in the area around the ground? Are views of local residents supported by your observed evidence?
- What is the extent of the effect of the club on the local economy? Do local residents and traders have similar views? Why or why not? Does age and gender affect residents' views?
- Summarise the impact of the football ground in the form of a cost benefit analysis (see figure 61, page 84). Identify the different groups affected and the social, economic and environmental issues involved.
- If the football ground is situated in an area of redevelopment use the planning proposals to assess the importance of the ground in those plans. Use all the information gathered from different groups, and an annotated map of the new plans, to note and assess the changes which are proposed for the area.

Resources

Goad street plan
Large-scale OS map
Website of football club
Local structure plan / development plan if appropriate
'In the Shadow of the Stadium: Football grounds as urban nuisances', John Bale, *Geography*, Volume 75, 1990

An assessment of a coastal management scheme

Starting Points ▶

1. What does the coastal management scheme involve?
2. Why was it necessary?
3. Does the scheme appear to be effective?
4. Do the costs outweigh the benefits?
5. What are the options for the future?

Geographical links to your syllabus

- Some sections of coastline are under increasing threat from erosion.

- Protection of the coast can be expensive as well as having detrimental effects on coasts elsewhere. Consequently the viability of coastal protection measures is increasingly being called into question.

Primary Data Collection

1. Make a detailed description of the current situation along the coast using annotated photographs and sketches. After your pilot survey, classify the types of erosion in the area. Look for evidence of former and perhaps continued erosion in the form of slumping down cliffs, cracks in cliffs, bare soil on the cliff face, surface water flowing over the cliff, rill patterns on the cliff face, undercutting of the base of slopes by wave action, rapid removal of material deposited on the beach. Note the type and amount of each piece of evidence. Create a sliding scale to represent the intensity of erosion of a particular process at a particular section of the coast (eg 3 = very severe, 1 = minor impact).
2. Survey the coastal zone from beach inland using surveying equipment. How steep are the cliffs? What is the height of land behind the immediate coastal zone?
3. Observe the wave frequency and height at different sites along the coast. Try to coordinate this with your observations on erosion. Although this will not tell you anything about the waves which cause erosion, it will give you some comparison between the different effects of waves at different points in your study area. If you can collect wave data in different wind conditions so much the better.
4. If there are groynes along the coast, measure the height of beach material from the top on each side. Look for the pattern on a series of groynes along the coast.
5. Look for evidence of intensive sea damage or erosion along the road, sea front and area behind the beach.
6. Make an index of visual quality (see figure 70) at different sites to establish whether the engineering has enhanced the coast or not.

7. Conduct a questionnaire of local residents to find out their views on local coastal protection initiatives. What damage has been done to their property? How frequently? Do they think the scheme is worthwhile? Effective? Sufficient? Would they pay more tax to fund more protection schemes?
8. Interview local businesses. How much of the business depends on the tourist trade? Are current management schemes protecting the tourist industry? Should there be more coastal protection? Would that prevent tourists from coming to the area?
9. Find out the land values of local property from newspapers. Interview estate agents to find out how long properties take to sell along the coastal strip. They may have experience of adverse surveyor's reports from houses in the area.
10. Interview the local planning officer and local engineers to identify the options for coastal management in the area.
11. Briefly observe the coastal erosion situation along adjacent sections of coast and assess the impact of your scheme on those areas.

⚠ *Things to look out for*

Your personal safety is important. Do not take risks. Use tide tables to make sure when it is safe to work. Check the weather forecasts.

This issue can be emotive. Be aware of the factors which influence the views of all the people you talk to.

Before you question anyone be fully informed about the area. Do the other practical work first so that you are quite sure what you need to find out.

Secondary Data Collection

1. Research information from local council engineering departments regarding the costs of the coastal management scheme, both predicted and actual, and the predicted lifetime of the engineering works.
2. Use evidence from local histories, old maps, old newspapers and old photographs to determine how the coast used to look.
3. Find out the current planning restrictions for the coastal zone.
4. Classify articles from the local press describing both the physical and human impacts of the coastal erosion. If you can, date the rate of erosion from these sources of evidence.

	Issue(s)	Summary of reaction and view point	Aspects of the issue which are of greatest concern to the interviewee
Name of interviewee			

Figure 72 Monitoring and summarising an interview

Ideas for recording and analysing data ▶▶

1. Draw an annotated map and sketch cross sections to describe the coastal management scheme(s).
2. Tabulate evidence of coastal erosion, past and present, at different sites along the coast.
3. Draw detailed profiles of beach / cliffs and the immediate area inland. Annotate to show the limit of storm damage. Integrate sketches or photgraphs of coastal erosion and protection.
4. Compare wave frequencies and wave height at different sites. Is there a relationship between sites with highest waves and sites which have the most evidence of erosion?
5. Draw a map or diagrams of the groynes to show how much beach material has accumulated.
6. Calculate the index of visual quality at various sites and, with the aid of photographs, assess how attractive the coast is now.
7. Summarise or graph the results of the questionnaires to local residents and businesses.
8. Tabulate articles and letters from the local press (see figure 65) concerning coastal erosion and the management scheme(s).
9. Use a similar outline to that above (figure 72) to record details of the interviews with the local planning officer and local engineers.

Interpretation and Conclusions

- Summarise the extent of former erosion along the coast and assess the extent of current erosion.
- What are the advantages and benefits of the scheme(s) for different groups of people?
- What has been the overall impact of the scheme over adjacent sections of coast? Has the scheme been cost effective? Summarise with a cost benefit analysis.
- Finally, your personal view based on the evidence you have accumulated – is coastal protection worth the money?

Resources

Local engineering reports of the coastal management scheme(s)
Records / leaflets from local Civic Trust or local history group

Analysis of the patterns of crime and deprivation in an urban area

Starting Points ▶

1. Some urban areas experience higher levels of crime than others.
2. The level of crime is associated with the social, economic and environmental characteristics of an area.
3. It is possible to quantify and map levels of deprivation and compare these with theoretical land use models.

Geographical links to your syllabus

- There are distinct socio-economic patterns in cities.

- The processes of decay, redevelopment, counter-urbanisation and industrial change all affect the quality of life in cities.

- Deprivation is an element of urban structure.

- For the general public, deprivation is part of the stereotype linked with the inner city, crime and related social problems.

⚠ Things to look out for

Your personal safety is important. Do not visit areas in the evening. Do not act suspiciously when recording your data.

Many crimes go unreported in some areas of towns where there are low rates of insurance cover, there is distrust of the police, and there is a poor understanding of community support and personal rights.

Criminal Statistics for England and Wales are available but only rates per 100,000 people. There is little available below that scale, so the main source of data will be the local press.

Primary Data Collection

1. Select 10 wards in your urban area using your local knowledge of environmental quality. ED areas are smaller and more homogeneous but you are unlikely to access some relevant data for those areas. Try to select areas which have distinct types of housing, eg detached houses, blocks of flats, nineteenth-century terraced housing, 1930s semi-detached houses, former council housing.
2. For each area assess the environmental quality (see figures 44, page 58, 63, page 86 and 64, page 88) based on a length of road. Use indicators such as: amount of graffitti, number of burglar alarms, number of cars parked in the road, quality of house maintenance, number of houses with double glazing, amount of open space, housing density, noise, vandalism and broken glass.
Give each indicator a score from 1 = worst / undesirable to 5 = best / very desirable. Total the score for each road and identify the highest and lowest scoring wards. These could be the focus of more detailed study if you wished to focus on building design and private space.
3. Collect details of house prices for each area. Be aware that these prices are not truly comparable if the house type is different.
4. Observe evidence of home security such as gates, entry phones to flats, burglar alarms, double glazing, security lights. Note also the extent of Neighbourhood Watch schemes.

Secondary Data Collection

1. Use the 1991 Census to collate socio-economic data for each ward on: levels of unemployment, types of employment, home ownership, housing density, housing quality, population structure, ethnic group, and mobility of population in each area.
2. Refer to local newspapers over a period of time to record and classify incidences of crime and convictions in each of your study areas.
3. Consult the local housing department to find out the policy for managing housing estates.
4. Collect details of local Neighbourhood Watch schemes from the community police.

Ideas for recording and analysing data ▶▶

1. Draw a choropleth map of each factor you have studied from the Census data, and also the environmental quality survey. Annotate to outline the general pattern and any anomalies you observe.
2. Draw scatter graphs to investigate interrelationships between sets of data, eg number of crimes in a ward *v* level of unemployment. Remember that there must be some logic to the links you identify between one set of data and another. Just because two sets of data

correlate does not mean cause and effect. Correlate using Spearman Rank and test for significance.
3. Apply an index of dissimilarity to compare the distribution of some of the variables you have mapped, such as different employment categories, different ethnic groups or the pattern of different house types (see Lenon and Cleves, page 104).
4. Calculate the Z scores for each ward in your study (see figure 42, page 56).

Interpretation and Conclusions

- Describe the patterns of environmental quality and socio-economic indicators throughout the urban area. Are the patterns of each indicator the same? Are they related to type of housing?
- Does the type of crime vary in different parts of the urban area? What influences the pattern of crime? Consider environmental factors including your observations in each street. How effective are Neighbourhood Watch schemes?
- To what extent are the patterns of crime related to patterns of deprivation?
- To what extent can you apply the theoretical urban land use models to explain your results?

Resources

Census 1991
Local newpapers
'The Challenge of Urban Crime' by D Smith, *Social Problems in the City* edited by D Herbert and D Smith, pp271–88, Oxford University Press, 1988
The Geography of Crime by D Evans and D Herbert, Routledge, 1982
'Urban Crime' by Garrett Nagle, *Geographical Magazine*, March, 1995
'Deprivation and the 1991 Census' by Robin Holmes, *Geography Review*, Volume 8, Number 3, January 1995
'The Geography of Crime and Crime Prevention' by Richard Yarwood, *Geography Review*, Volume 9, Number 3, March 1996
'Approaches to investigating the geography of crime', *Teaching Geography*, Volume 22, Number 1, January 1998
Crime and Disorder Audit, and Community Safety Strategy – both available from local police forces and local authorities

An assessment of the environmental impact of a landfill site

Starting Points ▶

1. How is the landfill site managed?
2. What are the issues of concern for local residents, including farmers?
3. How does the landfill site affect the local environment?

Geographical links to your syllabus

- Population growth is increasing pressure on natural resources.

- There is greater awareness of the need for sustainable management of the natural environment.

⚠ *Things to look out for*

You will find it easier to record noise levels on a calm day when there is less turbulence.

Don't approach this study in a negative way. Landfill sites are essential and operations are generally conducted with a high level of regard for the environment.

Primary Data Collection

1. Map a wide area around the landfill site and establish sampling sites on a systematic basis.
2. Measure noise levels at each sampling point.
3. On main roads leading to the landfill site survey the type and amount of traffic at regular intervals during the day and week.
4. Attach baby wipes to hedges or lamp-posts on the approach roads to the site for one week to record the amount of dust in the environment.
5. Measure soil pH of samples from the land surrounding the landfill site, and at 500m and 1km distance from the boundary.
6. Similarly if there are local streams, take water samples at sites flowing in and out of the landfill area to test for acidity and chemical content such as ammonia. If the streams appear murky use a Secchi disc (see figure 6) to identify changes in turbidity. A kick test (see figure 15) will enable you to test for pollution.
7. Conduct a questionnaire of local residents to identify their views on the landfill sites. How long have they lived in their house? Are they disturbed by noise, traffic or smell from the landfill site? Has the landfill site affected property prices?
8. Interview the manager or public relations officer for the landfill site. Why was this site chosen? Are there specific physical charcteristics which make it particularly suitable for its purpose? How is the plant managed? What environmental protection strategies operate there? How effective does the company think they are? What is the lifetime of the site? What strategies are there for returning the land to its former use?

Secondary Data Collection

1. Obtain details of the landfill operation from the operating company.
2. If the landfill site has opened fairly recently the local planning office will have details of the planning application.
3. Collect data from the environmental monitoring undertaken by the council.

Ideas for recording and analysing data ▶▶

1. Draw a site map of the landfill to show details of topography and drainage, and surrounding land use.
2. Draw a flow line map to show traffic flow on approach roads to the landfill site.
3. Classify the dust or greyness on the babywipes on a sliding scale, eg from 1 = evidence of large amounts of dust in the air to 5 = no evidence of dust in the air. If the dust levels decrease wih distance from the site draw an isoline map to show changes. Draw a similar map to show the distribution of noise levels and the pH results of soils.
4. Graph the stream pH levels for sites flowing in and out

of the landfill area to identify any differences.
5. Draw a large-scale map centring on the landfill site and make an annotated summary of the results of your observations.
6. Graph the responses of your questionnaire and compare with the information from the landfill company and the details provided in the planning application submission. Identify the main issues of concern if there are any.
7. Tabulate the results of your interview with the manager of the company.

Interpretation and Conclusions

- How does the landfill site affect the local environment?
- Make a cost benefit analysis of the landfill operation (see figure 61, page 84).
- Is the company managing the site effectively?
- Are the concerns of local residents significant?

Resources

Website of the landfill company
Local authority waste disposal plan
'Waste Management: Landfill or incineration?', *Geography Review*, Volume 8, Number 4, March 1995

Useful general resources

Texts

A–Z Geography Handbook, M Skinner, D Redfern and G Farmer, Hodder & Stoughton, 1999
Advanced Geography Fieldwork, J Frew, Nelson, 1993
Field Techniques and Research Methods in Geography, R H Stoddard, Kendall Hunt, 1992
Fieldwork Firsthand, Peter Glynn, Crakehill Press, 1998
Fieldwork Techniques and Projects in Geography, B Lenon and P Cleves, Collins, 1994
Geographical Data, H Matthews and I Foster, Oxford University Press, 1989
Geographical Data Analysis, N Walford, Wiley 1985
Geographical Techniques, Liz Taylor, Pearson Publishing, 1997
Methods of Presenting Fieldwork Data, Peter St John and Dave Richardson, Geographical Association, 1997
Methods of Statistical Analysis of Fieldwork Data, Peter St John and Dave Richardson, Geographical Association, 1996
Skills and Techniques for Geography A-Level, G Nagle with M Witherick, Stanley Thornes EPIC series, 1998

Geography Review

'De-stressing Statistics', Volume 11, Number 5, May 1998
'Environmental Impact Assessment', Volume 12, Number 2, November 1998
'Exploring Secondary Data', Volume 12, Number 5, May 1999

General websites

British Government information: www.open.gov.uk
Business Information Zone: www.thebiz.co.uk/categories_results
Countryside Council for Wales: www.ccw.gov.uk
English Nature: www.english_nature.org.uk
Environment Agency: www.environment-agency.gov.uk
Field Studies Council: www.field-studies-council.org
Local authorities, eg Cheshire County Council: www.cheshire.gov.uk
RSPB: www.rspb.org.uk
UK Office for National Statistics: www.ons.gov.uk
Maps of places in the UK: www.streetmap.co.uk/streetmap.dll

Index